THE AI RENAISSANCE

HOW A PROGRAMMER, A CEO, AND A PROFESSOR BUILT THE FOUNDATION FOR MODERN AI

SENTA BELAY

THE AI RENAISSANCE

How a Programmer, a CEO, and a Professor Built the Foundation for Modern AI

From Word Models to World Models

Written by Senta Belay

Published by Podcast Digests LLC Arlington, Virginia

ISBN: 979-8-9946265-0-4 (ebook) - $9.99

ISBN: 979-8-9946265-1-1 (paperback) - $17.99

ISBN: 979-8-9946265-3-5 (hardcover) - $29.99

ISBN: 979-8-9946265-2-8 (audiobook) - $14.99

First Edition: February 2026

CONTENTS

PREFACE

Three years ago, ChatGPT launched.

Four years is how long it takes to earn a degree. You pick a field, you study it deeply, you emerge with expertise that's supposed to last a career.

But here's the thing: any degree you started the day ChatGPT launched is already, in some fundamental way, obsolete. Not worthless, but incomplete. The world you trained to enter no longer exists. The tools you learned to use have been superseded. The problems you prepared to solve have been transformed.

Three years. That's all it took.

This book exists because I needed it and couldn't find it.

When I first felt the impact of AI becoming real (not as a headline, not as a demo, but as something that changed how I worked and thought) I wanted to understand how we got here. I wanted to know who built this, and why, and what they were thinking. I wanted the story.

What I found was fragments. Technical papers. Podcast clips. Business news. Conference talks. The pieces of a story scattered across the internet, told by people who rarely talked to each other, understood by almost no one.

So I wrote the book I wished I had.

The first thing to understand: this didn't happen overnight.

The threads of this revolution stretch back decades. A programmer in Texas who wanted to make video game demons run faster. A CEO in California who nearly went bankrupt three times before his company became the most valuable on Earth. A professor in Canada who kept faith in neural networks through forty years when the field declared them dead.

They never met while building the same thing. They had no idea they were constructing the same cathedral. A demand for faster graphics. A platform for parallel computing. A theory of how machines could learn. Three obsessions that converged, accidentally, into the foundation for artificial intelligence.

The story of how we got here unfolds slowly. It took thirty years. It required patience and luck and survival through multiple collapses. And then, once the foundation was built, everything went into overdrive – the scaling, the money, the arms race between companies and countries, and finally the doubts about whether scaling alone would be enough.

This book tells you who they are.

You'll meet them as human beings: flawed, uncertain, brilliant, sometimes frightened of what they've built. The godfather of deep learning who won the Nobel Prize and then left Google so he could warn the world. The CEO who still wakes up afraid, every morning, despite running a four-trillion-dollar company. The programmer who made demons run faster and is now trying to make machines actually think.

You'll learn what they did, what they're doing now, and most importantly: how they think and what they pursue.

And you'll see where this story is heading now – from **word models to World models** – as a new cast of builders converges, once again from different directions, on the deepest question: what intelligence actually is, and what it could become.

But here's what makes this moment different: the godfathers themselves no longer agree on the answer. Some believe the future lies in world models – AI that understands physics, not just language. Others, like Yoshua Bengio in Montreal, have concluded that the problem runs deeper: any system that pursues goals will eventually pursue the wrong ones. His answer? Don't build better agents. Build **Scientist AI** – intelligence that predicts without wanting, understands without goals, watches over other AI like the laws of physics themselves.

The conspiracy that built this revolution has fractured. The heretics have multiplied. And the future may require both approaches – or neither.

By the time you read this, something will have changed. The people you'll meet in these pages are still working, still arguing, still pushing toward destinations none of them can clearly see.

This story has no ending. It can't.

I hope this book helps you find your place in it.

Senta

February 2026

INTRODUCTION: THE CATHEDRAL BUILDERS

JOHN CARMACK, EARLY 20s
JENSEN HUANG, EARLY 30s
Denny's
GEOFFREY HINTON, 40s

This is a story about three people who never met while building the same thing.

The first was a programmer in Texas who wanted to make demons run faster on computer screens. The second was a CEO in California who bet his company's future on a technology nobody wanted. The third was a professor in Toronto who spent forty years on an idea most of his colleagues had abandoned.

None of them were trying to create artificial intelligence.

Yet their separate obsessions, pursued independently across different decades for entirely different reasons, converged to build the foundation for the most transformative technology of our era.

They were building the same cathedral. None of them knew it.

This is the story of how that happened – and why it matters for what you might build next.

THE THREE

John Carmack

In 1993, a twenty-three-year-old programmer in Mesquite, Texas released a video game called Doom. It spread across the nascent internet like a virus. Millions downloaded it. Every one of them wanted better hardware to run it.

Carmack wasn't thinking about AI. He was thinking about how to render Hell in real-time. His solution was to break complex calculations into thousands of parallel operations, a technique that would accidentally establish the computational philosophy powering neural networks.

By 1996, his next game, Quake, would force an entire industry into existence. Graphics cards went from luxury to necessity. The demand Carmack created would fund GPU development for three decades.

He made demons run faster. Everything else followed.

Jensen Huang

In 1993, the same year Doom shipped, a thirty-year-old engineer sketched a business plan on napkins at a Denny's restaurant. He and two co-founders would build chips to satisfy the hunger Carmack had created.

The company was called Nvidia. It nearly died twice before Jensen Huang made the bet that would matter: over a billion dollars on CUDA, a technology that would let scientists use gaming chips for research.

For years, almost nobody cared. Jensen kept investing anyway, convinced that parallel processing would matter for something beyond games. He didn't know what. He was building infrastructure for a future he couldn't see.

Then the researchers found him.

Geoffrey Hinton

In 2012, nineteen years after Doom, six years after CUDA, a British professor at the University of Toronto entered his students in an image recognition competition.

Geoffrey Hinton was sixty-four years old. He had been working on neural networks for over forty years, through two "AI winters" when the field declared his approach dead. He had moved to Canada because the funding was better and the skepticism was softer. He had trained a small army of graduate students in techniques nobody else believed in.

His students built a neural network called AlexNet. They trained it on two graphics cards, gaming hardware at $500 each. They entered the ImageNet competition expecting a reasonable performance.

They didn't just win. They annihilated the field. Their error rate was 15.3%. The next best entry was 26.2%. It wasn't a competition. It was a revelation.

The hardware that Carmack's games had demanded, that Jensen's company had built, had just proven Hinton's forty-year bet was right.

The threads had crossed.

THE THESIS

We are in the midst of a great technological renaissance.

Large language models have brought a renaissance of thought to the masses. The tools now enable action at the speed of thought; the critical friction is no longer execution, but whether you're thinking about the right things.

This transformation extends beyond text: voice, video, images, and eventually "world models" that replicate full human experience. The book documents both the promise (democratized creation, accelerated science, new forms of expression) and the costs (hardware scarcity, job displacement, regulatory gaps, concentration of power).

But before any of that, three people had to build the foundation.

A programmer who made demons run faster.

A CEO who survived long enough to matter.

A professor who kept faith when no one believed.

The same cathedral.

Now you know.

THE ARC

This story unfolds in four acts.

ACT I: BUILDING THE CATHEDRAL

The Origin Stories (1988–2012)

Three people working on problems that seemed unrelated, unwittingly constructing the infrastructure for machine intelligence.

The Programmer. John Carmack, 1988–1999. How a juvenile delinquent became the most important programmer of his generation. How Doom created a market and Quake made it permanent. How the demand for graphics hardware was born.

The CEO. Jensen Huang, 1963–2006. How an immigrant child who cleaned toilets in rural Kentucky built a company that almost failed twice. How Nvidia survived the graphics card wars. How the CUDA bet was made, and why nobody noticed.

The Professor. Geoffrey Hinton, 1947–2012. How a believer kept faith through forty years of rejection. How neural networks survived two AI winters. How a small lab in Toronto prepared for a breakthrough that might never come.

The Convergence. October 2012. The moment the threads cross.

AlexNet wins ImageNet. The gaming hardware Jensen built, funded by the demand Carmack created, proves Hinton's life work was right.

Interlude: The Coffee Shop. How neural networks learn. Backpropagation explained. The mathematics of blame.

The Builders, Afterward. The three who built the foundation step back. Where Carmack, Jensen, and Hinton went next. The explicit handoff to the scalers who would occupy the cathedral.

ACT II: THE EXPLOSION

The Decade of Scale (2012–2022)

The accidental infrastructure works. Deep learning goes from curiosity to dominant paradigm.

The Scalers. The talent wars begin. Google buys Hinton's company. The race for researchers reshapes Silicon Valley.

The Foundations. AlphaGo's Move 37. The scaling hypothesis crystallizes. Transformers arrive. GPT proves language models can be general-purpose.

The Interface. November 2022. ChatGPT launches. One hundred million users in two months. The fastest adoption in technology history. AI stops being abstract and starts being real.

Interlude: The Library. How language models work. Embeddings, attention, and the Transformer architecture.

Interlude: The Teacher. Andrej Karpathy explains how it all fits together. The pause before the race begins.

ACT III: THE SCALING FAITH

The Race (2023–2024)

ChatGPT proved the cathedral matters. Now everyone races to control it. The scaling laws become an article of faith.

The Kingmaker. Jensen's payoff arrives. The GPU shortage. Nvidia's valuation explodes. The foundation layer is controlled.

The Faith. Why everyone believed. The scaling laws. The bitter lesson. Loss curves that didn't bend. OpenAI, Anthropic, Google, Meta: different strategies, same underlying conviction.

The Fracturing. The believers divide. Safety versus speed. Open versus closed. The OpenAI board crisis. Anthropic's founding. The talent wars. Hinton walks away.

The Eastern Front. China enters the race. DeepSeek proves efficiency can challenge scale. The study group graduates. A different path emerges.

Interlude: The Game. How networks learn from doing, not reading. Reinforcement learning explained. The alternative paradigm before Act IV explores the researchers pursuing it.

ACT IV: THE HERETICS RETURN

The Questioning (2024–?)

The builders themselves start questioning. The prophets of scaling announce limits. The case that prediction isn't enough.

The Doubters. Ilya's "peak data" speech. The prophet of scaling announces the fuel is running out. Hinton's warnings from outside. The Turing Award winners divided. The builders' uncertainty.

The World Builders. The alternative thesis emerges. Rich Sutton's critique: the squirrel test. Carmack partners with the father of reinforcement learning. Fei-Fei Li's spatial intelligence. LeCun's world models bet. The case that language models are "ghosts, not animals."

The Cathedral. Return to the three protagonists. Carmack's view of what he contributed. Jensen's "thirty days from bankruptcy" at a $4 trillion valuation. Hinton's regret, warning, and pride. The cathedral incomplete. The builders arguing about blueprints for the next wing.

Four interludes throughout the book explain the technology as it becomes relevant to the story. Even if you already understand neural networks or transformers, the interludes connect what the technology does to why the builders made the choices they did. They're not departures from the narrative – they're the view from inside the machine.

ACT I: BUILDING THE CATHEDRAL

THE ORIGIN STORIES (1988–2012)

Three people who never met, working on problems that seemed unrelated, laid the foundation for machine intelligence without knowing it.

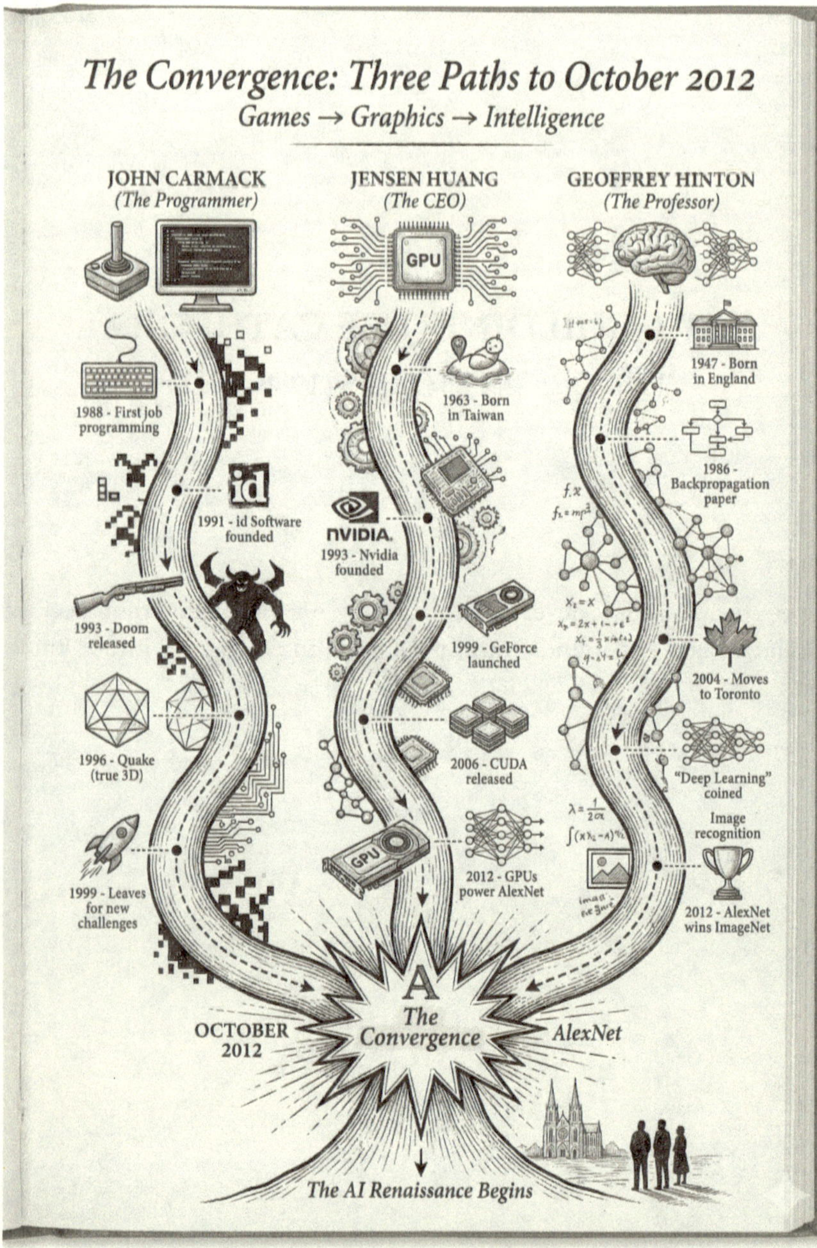
The Convergence: Three Paths to October 2012
Games → Graphics → Intelligence
JOHN CARMACK
(The Programmer)
JENSEN HUANG
(The CEO)
GEOFFREY HINTON
(The Professor)
GPU
1988 - First job programming
1991 - id Software founded
1993 - Doom released
1996 - Quake (true 3D)
1999 - Leaves for new challenges
1963 - Born in Taiwan
NVIDIA
1993 - Nvidia founded
1999 - GeForce launched
2006 - CUDA released
GPU
2012 - GPUs power AlexNet
1947 - Born in England
1986 - Backpropagation paper
2004 - Moves to Toronto
"Deep Learning" coined
Image recognition
2012 - AlexNet wins ImageNet
A
The Convergence
OCTOBER 2012
AlexNet
The AI Renaissance Begins

1

THE DOOM EQUATION

The Connection Nobody Saw

A programmer in Texas made demons run faster.

That's it. That's the origin of everything.

John Carmack, twenty-three years old, wanted his video game to feel *real* – visceral, immediate, terrifying. The computers of 1993 couldn't deliver what he saw in his head. So he figured out how to break complex calculations into thousands of parallel operations, running them simultaneously rather than sequentially.

He was solving a problem in front of him. He had no idea he was establishing the computational architecture for machine learning.

The same parallel processing that renders demons also trains neural networks. When a neural network learns, it performs millions of simple mathematical operations simultaneously. Each individual operation is trivial. But there are so many of them, and they can all happen at once, that sequential processing becomes impossible.

The kid playing *Doom* in 1994, complaining that the frame rate stuttered when too many demons appeared on screen, was creating market pressure for exactly the hardware that would eventually train ChatGPT.

This book is about how that happened. How tools for play became tools for thought. How the nerds didn't just win culturally; they built the foundation for thinking machines without realizing it.

The Dungeon Master's Dice

Before video games, there was Dungeons & Dragons.

The game emerged in 1974 from a simple innovation: instead of commanding armies across hex grids, each player controlled a single character. A fighter. A wizard. A thief. When players attempted actions not covered by the rules, the Dungeon Master adjudicated on the fly. This broke the closed loop of board games, where only codified moves are legal. It opened the possibility of *anything*.

D&D taught millions of young people to think in systems. Hit points. Armor class. Experience levels. Saving throws. Before video games existed in any sophisticated form, tabletop role-playing established the grammar of interactive fiction: you have a character, the character has statistics, the statistics determine outcomes, and the outcomes unfold as story.

Every video game designer who came of age in the 1970s and 1980s played D&D. Every single one. When they started building digital worlds, they built them in the language Gary Gygax and Dave Arneson had invented. Health bars are hit points. Character progression is experience levels. The entire architecture of interactive entertainment descends from a game played with paper, pencils, and polyhedral dice.

The worlds those dice conjured demanded better hardware – faster rendering, richer environments, more complex simulations. That demand, through a chain nobody planned, produced the GPUs that would eventually train ChatGPT.

Masters of Doom

Masters of Doom, David Kushner's 2003 chronicle of id Software, tells the story of two Johns – John Carmack and John Romero – who created the games that defined a generation: *Commander Keen*, *Wolfenstein 3D*, *Doom*, *Quake*.

Kushner spent months reporting. He visited the buildings where id Software had worked, interviewed everyone who was there, triangulated conflicting accounts of dramatic moments. "Every line of dialogue, every scene, every detail is drawn from reporting," he insists. "I take zero artistic license."

The book didn't just document history; it shaped future history. Palmer Luckey read it and was inspired to found Oculus VR, where he eventually hired Carmack himself. An entire generation of tech entrepreneurs saw themselves in the Two Johns.

Doom was more than a game. It was a proof of concept for parallel computing on consumer hardware.

The Programmer

To understand what Carmack did, you have to understand who Carmack was.

The love affair with computers started before he ever touched one. He remembers looking through old encyclopedias at black-and-white photos of IBM mainframes, feeling something pull at him. He would go to libraries and devour whatever few books on computing he could find.

"It felt like this magical thing to me," Carmack has said. "This idea that the computer would just do exactly what you told it to. It wasn't going to back talk you. It wasn't going to have a different point of view. It was going to carry out what you told it to do."

His first program was at a Radio Shack, on a TRS-80 display model: 10 PRINT JOHN CARMACK. Then 20 GOTO 10. His name scrolling endlessly down the screen.

The childhood that produced this obsession was not easy. High school was misery. "I was unhappy most of my teenage years," Carmack has reflected. "I really wanted to be on my own doing programming all the time." When the school counselor ran through the standard script for troubled kids, asking if he ever thought about ending it all, Carmack's answer was certain: "This is temporary. Things are going to be better."

What saved him were the escapes: books, comic books, Dungeons & Dragons, arcade games. His fondest childhood memories are convenience stores–7-Elevens and QuikTrips – because they had spinner racks of comics and little side rooms with arcade cabinets. That was his happy place.

The moment he could support himself programming, even poorly, even barely, he was happy. "As soon as I was 18, 19, even though I was poor, I was doing exactly what I wanted."

The work *is* the reward. He has never experienced burnout. Not once. "I've never been at a point where I'm like, I just don't care about this." When stress hits, he copes by bearing down harder. "I pull myself out of whatever hole I might be slipping into by actually making progress."

People who work with him describe someone who appears grim: knitted brows, a frown of concentration. Carmack acknowledges this. "That's what happiness looks like for me."

Sixty hours a week, for decades. This was the person sitting in a Mesquite, Texas office in 1993, staring at lines of code, trying to make demons run faster.

The Problem

Rendering a three-dimensional world on a two-dimensional screen requires enormous amounts of calculation.

Which surfaces are visible from the player's current position? How does light fall across those surfaces? What color should each individual pixel be, given the textures and the lighting and the distance?

Every frame requires answering these questions for hundreds of thousands of pixels. And Carmack wanted thirty-five frames per second. That meant answering millions of questions every second, continuously, without any perceptible delay.

The CPUs of 1993 processed instructions sequentially. One calculation, then the next, then the next. This approach was brilliant for spreadsheets and word processors, tasks where each step depends on the previous step, where order matters.

It was terrible for rendering Hell in real-time.

The Restaurant Kitchen

Think of a CPU as a single, brilliant chef.

Complex tasks, one at a time. *Simple tasks, all at once.*

This chef can prepare any dish. Sushi. Soufflé. Tandoori chicken. Molecular gastronomy. She's extraordinarily skilled and extraordinarily fast. She has decades of training and an encyclopedic knowledge of techniques. Give her any recipe, no matter how complex, and she'll execute it flawlessly.

But she works alone. One dish at a time. Start to finish.

Now imagine you need to serve 35 seven-course meals per second.

No chef, however brilliant, can work that fast. The sequential approach breaks down when the quantity exceeds what any single processor, human or silicon, can handle.

Carmack's approach was to fire the brilliant chef and hire a thousand line cooks.

Each one could only do one thing. This one chops carrots. That one flips patties. Another one boils water. No creativity. No flexibility. No ability to handle unexpected requests or adapt on the fly. Each cook knows exactly one operation and performs it endlessly.

But they all work simultaneously.

You need a thousand carrots chopped? All thousand cooks chop one carrot each. Done in the time it takes to chop one.

The final meal materializes not through sequential excellence but through parallel adequacy. No individual cook is impressive. The collective output is miraculous.

This trade-off – less capability per unit, massively more units working in parallel – is the fundamental architectural choice that defines the GPU.

Carmack didn't invent parallel processing. Supercomputers had used it for decades. Academic computer scientists understood the theory. The concept wasn't new.

What Carmack did was prove it could work on consumer hardware, for consumer entertainment, at consumer prices.

And he proved people would pay for it.

The Shareware Revolution

Doom shipped on December 10, 1993.

The distribution method was unusual for its time. id Software released the first episode, nine levels, for free. Anyone could download it from bulletin board systems, the pre-internet networks where enthusiasts shared files by dialing in over phone lines. Copy it. Share it with friends. No charge.

If you wanted more – the full game, all thirty-six levels, the complete journey through Hell – you paid thirty dollars, sent directly to id Software in Texas.

Within weeks, *Doom* had spread across every bulletin board in America. Within months, it had been downloaded millions of times. Within a year, hundreds of thousands had paid for the registered version.

id Software, with a staff you could count on your fingers, was earning six figures per month.

The money was astonishing. But here's what matters for our story: *Doom* created a market.

Millions of players suddenly wanted better graphics. Faster frame rates. More detailed textures. Smoother animations. They wanted hardware that could run *Doom* better, and run the increasingly demanding games that followed.

Doom begat *Quake*. *Quake* begat *Half-Life*. Each generation pushed hardware harder, demanded more parallel processing power, rewarded companies that could deliver it.

The demand Carmack created would fund GPU development for the next decade.

Neural networks are what computer scientists call "embarrassingly parallel," a technical term that means the calculations don't depend on each other. You don't need to finish one before starting the next. They can all happen simultaneously.

Exactly like rendering pixels in a video game.

The demons running across screens in 1994 were ancestors of the models that would write poetry and pass bar exams and generate images from text descriptions.

And it started, like everything else, with a programmer who had a problem to solve.

Demons to make faster.

2

THE QUAKE REVOLUTION

The Metaverse Before There Was a Name

The Metaverse came from a science fiction novel. But the first serious attempt to build it happened in a game studio in Texas.

Neal Stephenson's *Snow Crash*, published in 1992, described the Metaverse – a virtual reality space where millions of people gathered as avatars, navigating digital streets, entering virtual buildings, experiencing things impossible in the physical world. The novel was prescient in ways that would take three decades to fully appreciate. Mark Zuckerberg would eventually rename his entire company after the concept.

But here's what most people don't know: the first serious attempt to build that vision wasn't made by a Silicon Valley giant. It was made by a small game company in Texas, and the conversations that shaped it happened between two programmers who had both read Stephenson's novel.

"We had metaverse talk back then," John Carmack recalls. "We knew about this. We would talk about all that and kind of spin this tale."

The conversations happened during the development of Quake. Carmack and Michael Abrash, the legendary assembly language wizard Carmack had recruited from Microsoft, would discuss what games could become. "Different areas linked together by portals," Carmack remembers them imagining. "User-generated content. Changing out all of these things."

They were describing social media. They were describing Second Life. They were describing the vision that would eventually consume Facebook.

"It was some of the same conversations that we have today about the metaverse," Carmack says.

This was 1995. The World Wide Web was barely two years old. Most people didn't have email addresses yet. And two programmers in Mesquite, Texas, were already imagining a future that venture capitalists wouldn't discover for another twenty-five years.

They just had to build it first.

The Magic Trick

Doom was an illusion.

The demons that hunted you through corridors, the shotgun that kicked in your hands, the walls that dripped with hellish texture – all of it was an elaborate conjuring act. You couldn't look up. You couldn't look down. The "3D" world was actually flat, a two-dimensional map extruded upward, cleverly disguised by Carmack's rendering tricks.

The technical term was "2.5D." You moved through a world that *felt* three-dimensional but was mathematically constrained. You couldn't have a room above another room. You couldn't tilt your head. The ceiling and floor were always at fixed heights relative to your viewpoint.

"Doom was still not a true 3D," Carmack has explained. "You did not have the ability to tilt your head this way or pitch up and down."

This was fine. More than fine: it was revolutionary. Doom had proven that first-person immersion was possible on consumer hardware. It had created a market. It had made id Software rich.

But Carmack knew it was a trick. And tricks have limits.

Quake would be different. Quake would be real.

The name itself came from Carmack's imagination, or rather from a Dungeons & Dragons campaign he ran in the early 1990s. "Quake" was a character in the "Silver Shadow Band," described as "the strongest and most dangerous person on the continent," wielding the Hammer of Thunderbolts. The team had originally envisioned a role-playing game

(RPG) built around this character: puzzles, narrative depth, a ring of regeneration.

That dream would die. But the name survived.

Biting Off More Than Could Be Chewed

The ambitions for Quake were staggering.

"It was going to be really a true 3D engine where it could do six degree of freedom," Carmack recalls. "You could have all the viewpoints. You could model anything."

Six degrees of freedom. Move forward, backward, left, right, up, down. Pitch, yaw, roll. Full navigation through genuine three-dimensional space. Not the constrained corridors of Doom, but actual geometry: rooms above rooms, surfaces at any angle, a camera that could point anywhere.

That alone would have been ambitious enough. But id Software piled on more.

A revolutionary lighting system. "Surface caching," Carmack called it – techniques "that they weren't doing even on the very high-end systems." Professional Silicon Graphics workstations, the kind that cost more than a house, weren't attempting what Carmack wanted to run on a consumer PC.

A completely programmable game engine. Doom had let players create new levels and swap out artwork, but the game logic was locked in compiled code. Quake would have its own programming language, QuakeC, that would let modders change *anything*. These were the hobbyists who took commercial games and made them their own – new levels, new weapons, entire new experiences built on borrowed foundations. "Code was going to be data," Carmack explains. "You would be able to have expansion packs that changed fundamental things."

And multiplayer over the internet. Not the peer-to-peer connections of Doom, limited by phone lines and local networks, but true client-server architecture that could support "larger numbers of players in disparate locations."

"We bit off a lot," Carmack admits now.

The RPG mechanics weren't working. Melee combat with the Hammer of Thunderbolts felt sluggish compared to Doom's instant gratification. The Aztec art assets the team had spent nearly a year creating sat unused, inspiring no one. In November 1995, American McGee ran an experiment: he copied a bunch of Doom textures into the Quake engine and built a quick test level.

It felt right.

"The idea of falling back to 'just' making a shooter didn't seem so bad," Carmack recalled. The Hammer of Thunderbolts was replaced by a shotgun. The RPG died. Quake lived.

Looking back, Carmack would still do things differently. "If I could go back and tell younger me to do something differently, it would have been to split those innovations up into two phases in two separate games." Keep the Doom graphics engine, add the networking and the programming language. Ship that eight months faster. Then tackle true 3D in the next game.

"We could have made more people happier and gotten two games done in 50 percent more time."

But that's hindsight. In 1995, they went for everything at once.

The First Limitation

For the first time in his career, John Carmack hit a wall.

"Quake was kind of the first thing where I did have to kind of come face to face with my limitations," he has said. "It was the first thing where I really did kind of give it my all and still come up a little bit short in terms of what and when I wanted to get it done."

This is a telling admission from a programmer whose entire identity was built on making the impossible possible. Commander Keen's smooth scrolling. Wolfenstein's ray-casting. Doom's binary space partitioning. Every previous project, Carmack had found a way through. Technical obstacles existed to be demolished.

Quake was different. Quake fought back.

"The company ran had some serious stresses through the whole project," Carmack recalls. The graphics engine "went a long time internally really not working." Level designers would build content only to watch it break as Carmack rewrote fundamental systems. "Large fractions of the world disappearing, not being there, or being really slow in various parts of it."

The timeline stretched. The team ground themselves down.

And Carmack kept working, trusting that the breakthrough would come.

"It was this act of faith," he says. "I think I'm going to be able to fix this. I think I'm going to be able to make this work."

How Carmack Survives

Here's something worth understanding about John Carmack: he doesn't break.

Not in the way most people imagine, anyway. When asked about dark moments during Quake's development, the stress, the impossible deadlines, the technical challenges that seemed insurmountable, Carmack's response reveals the architecture of his psychology.

"I've never really had really dark black places," he says. "I understand a lot of people have significant challenges with kind of their mental health and well-being. I've been super stressed, I've been unhappy as a teenager in various ways, but I've never really gone to a very dark place."

His coping mechanism is work itself.

"I've always coped best on that when all I need to do is usually just kind of bear down on my work," he explains. "I pull myself out of whatever hole I might be slipping into by actually making progress."

The logic is circular but functional: when things feel impossible, work harder. Working harder produces progress. Progress produces hope. Hope sustains more work.

"I always know if I just keep pushing, eventually I break through and I make progress. I feel good about what I'm doing."

This isn't advice. It's a description of unusual psychological architecture. Most people don't work this way. Carmack does, and it's part of why Quake got finished.

"I've borne up well under whatever stresses have kind of fallen on me," he says, with the tone of someone describing a trait like height or eye color: something he happened to be born with.

The Wizard from Microsoft

Carmack had a secret weapon: Michael Abrash.

Back in the Softdisk days, before id Software existed, Carmack had discovered Abrash's articles in *Dr. Dobb's Journal* – technical deep-dives into assembly language optimization that read like sacred texts. Abrash was legendary in graphics programming circles, the kind of coder who could squeeze impossible performance from recalcitrant hardware.

"Some of my most treasured ones were Michael Abrash's articles in *Dr. Dobb's Journal*," Carmack recalls. "He had been one of my early inspirations."

By the mid-1990s, Abrash held a senior technical role at Microsoft. He was on track for great things at a company that was about to dominate the technology industry. Microsoft was "starting to take off," Carmack notes. Abrash was positioned perfectly for the boom.

Carmack wanted him anyway.

"After all of our success in Doom we were able to kind of hit him up and say, 'Hey, we'd like you to come work at id Software,'" Carmack recalls. The pitch was audacious: leave Microsoft's certain trajectory for a small game company in Texas, because what they were building "was going to be something nobody had seen before."

Carmack and Abrash shared a vision. They had both read *Snow Crash*. They talked about the metaverse, user-generated content, virtual worlds that people would actually inhabit.

"We were talking about what these games could become," Carmack says. "We were seeing an explosion of work with Doom at that time where people were doing amazingly cool things – like we saw cooler levels than we had built coming out of the user community."

Quake wasn't just a game. It was infrastructure for something larger.

Abrash said yes.

Carmack considered Abrash one of the strongest programmers he knew – one of the few people he trusted to write assembly code better than himself. Coming from a self-described "world-class optimizer," this was the highest possible praise. Carmack made things go fast, but he recognized that a handful of people could write tighter assembly, tighter SIMD code than he could.

Abrash was one of those people.

The collaboration would prove essential, if sometimes painful. Abrash would spend a month crafting beautifully optimized assembly language for the core rendering routines. Then Carmack would discover a higher-level optimization – using a span buffer instead of painter's order, cutting 30% or 40% of pixels – that required rewriting Abrash's careful work.

It wore on Abrash, but the results justified the process. The final engine shouldn't have been possible on consumer hardware: a super-optimized assembly language core loop surrounded by a system that minimized how often it needed to be called.

The Fracture

The stress didn't just affect the code. It cracked the company.

John Romero, co-founder, co-visionary, the designer whose levels had defined Doom's feel, was struggling. The Quake development stretched into its second year, then approached its third. The team dynamic that had powered id Software's meteoric rise began to fail.

"I did push Romero out of the company," Carmack acknowledges now. Looking back, he sees how the original founding structure – equal partners, equal ownership – created problems under pressure. He was working harder than anyone, doing technologies nobody had done before, but they were all equal partners. When he saw someone not working as hard, he admits he wasn't mature about it. He was twenty-something years old.

The conflicts weren't just about hours logged. They were about competing visions for what id Software should become.

Romero wanted to expand, hire more people, run multiple teams simultaneously. Carmack resisted, valuing the small-team intensity that had produced their breakthroughs. He'd heard that companies turn to shit when they hit fifty employees. He loved their little dozen people working on projects.

The specific grievances seem almost quaint now. Romero not working as hard. "Not carrying his load on the design for Quake and coming up with things there," as Carmack put it.

Looking back, he sees it differently. "He was definitely doing things.

He made some of the best levels there. He was working with some of our external teams like Raven on the licensing side of things."

But in the moment, consumed by the impossible task of making Quake real, Carmack couldn't see past his own intensity.

"I know I was really, really angry," he admits.

What Romero Built

Here's what Carmack will now acknowledge: he didn't do it alone.

When Carmack first met Romero, he thought Romero was the coolest programmer he'd ever seen. Romero had made games at Origin Systems – one of the companies Carmack most admired – and he made things happen fast.

Romero was a polymath in a way Carmack wasn't. He drew his own art, designed his own levels, worked on sound systems, and was a genuinely good programmer. Early on, they even competed: a race to port a PC game back to the Apple II established that Carmack stood apart on the programming side. But Romero adapted gracefully, moving into tools, systems, and game design.

He was enormously valuable in the early years. So much of Doom and Quake bear his stamp.

Success brought its own complications. The press coverage about "rock star game programmers" and "game developers with Ferraris" changed the dynamic. It was the Beatles problem: fame fracturing friendship. Romero embraced the celebrity; Carmack thought it led to challenges.

After the split, Romero landed on his feet. Eidos gave him $20 million to start Ion Storm, where he spun up three teams simultaneously – exactly what he'd wanted at id. So much of what was great in the games came from him, and Carmack wouldn't take that away.

The Lennon and McCartney of video games had broken up. Unlike the Beatles, they'd ship one more album together.

June 22, 1996

Quake arrived.

For the first time on consumer hardware, players could look

anywhere. Up at ceilings arching overhead. Down into bottomless pits. At geometry that twisted and folded in ways Doom's architecture could never allow. The monsters weren't sprites anymore; they were actual 3D models, moving through actual 3D space.

The game's strange, nightmare-logic aesthetic – medieval castles connected to military bases, tentacled horrors lurking in impossible geometries – came from level designer Sandy Petersen, creator of the *Call of Cthulhu* tabletop RPG. Petersen filled the vacuum left by the abandoned fantasy RPG with cosmic horror. The final boss, Shub-Niggurath, was named directly after a Lovecraft deity. You couldn't kill it with weapons; you had to telefrag it, teleporting inside its body. Even victory felt like madness.

Trent Reznor of Nine Inch Nails created the soundtrack – not music, but "textures and ambiences and whirling machine noises." The NIN logo appeared stamped on ammunition boxes. Industrial music had merged with the code.

The graphics were "the showcase of what Quake was all about," as Carmack put it. Technology that hadn't been attempted even on expensive professional workstations was now running on PCs.

The cost was brutal. Quake required a Pentium processor, and even then "it was not a great experience" on lower-end machines. "We did cut out a lot of people with the hardware requirements for Quake," Carmack admits.

But for those who had the hardware, or who upgraded to get it, Quake was a revelation.

The Hardware Mandate

Here's what Doom had done: it had created *desire* for better graphics.

Here's what Quake did: it created *necessity*.

You could play Doom on a 486. It ran poorly, but it ran. You could experience the magic even on modest hardware.

Quake was different. Below a certain threshold, Quake was unplayable. The frame rate dropped to slideshow territory. The sophisticated lighting and real-time 3D calculations overwhelmed anything but the newest processors.

This was a problem for consumers in 1996. And it was an opportunity for an industry that barely existed.

Graphics cards, dedicated hardware for rendering images, had been a niche product. Expensive professional workstations used them. High-end CAD systems. Flight simulators for military training. But consumer PCs? They used whatever graphics capability was built into the motherboard.

Quake changed the equation. Suddenly, upgrading your graphics wasn't a luxury. It was the difference between playing the most anticipated game of the year and staring at a stuttering mess.

3dfx released the Voodoo card in late 1996. It was designed specifically for the kind of 3D acceleration Quake demanded. In January 1997, Carmack released GLQuake: a version of the game that could leverage OpenGL, the graphics programming interface that let software talk to dedicated hardware.

The difference was transformative. GLQuake on a Voodoo card ran faster and looked better than software rendering on the most powerful CPU available. The era of the graphics card had begun.

The OpenGL Wars

Carmack didn't just make games. He shaped the future of graphics programming.

His advocacy for OpenGL, against Microsoft's competing Direct3D, would become one of his proudest accomplishments, for reasons that took decades to become clear.

On Windows, it was always a rear guard action. Microsoft was never going to let an open standard win on their platform. But Carmack kept fighting anyway, writing technical arguments defending OpenGL's approach against what he saw as Microsoft's inferior alternative.

The payoff came when mobile devices arrived. Because Carmack and others had kept OpenGL relevant, when smartphones needed graphics acceleration, OpenGL ES was ready. The standard he'd championed became the foundation for the mobile graphics revolution.

Think about the scale. Carmack's games sold tens of millions of copies. But billions of devices ended up with capable graphics APIs because he thought open standards mattered.

The Android phones in everyone's pockets. The iPhones before Apple moved to Metal. The tablets, the smartwatches, the embedded systems. All of them descended from the standard Carmack championed in those early years.

"That would be on my greatest hits list of things that I kind of pushed back that had [impact] on billions of devices."

The Networking Mistake

QuakeWorld launched in December 1996.

But first, Carmack had to admit he'd been wrong.

He had developed Quake's original networking code on a high-speed T1 line in the office, assuming low latency and reliable connections. The code used TCP, a protocol that guaranteed delivery but introduced delays. On a corporate network, it worked fine.

On a dial-up modem, it was unplayable.

"I was working with the wrong basic assumptions for doing a good internet game," Carmack admitted in his .plan file. "My original design was targeted at <200ms connection latencies." Real-world players on 28.8k modems experienced far worse.

QuakeWorld was the fix. Carmack switched from TCP to UDP, a "fire and forget" protocol that didn't guarantee delivery. If a packet containing a player's position from 100 milliseconds ago got lost, the server didn't care. It only wanted the *current* position.

Then came client-side prediction. In the original Quake, when you pressed "forward," the command traveled to the server, the server moved your character, and the result came back to you. On a 300-millisecond ping, you waited a third of a second just to take a step.

QuakeWorld changed the architecture. When you pressed "forward," your local machine moved you immediately, *assuming* the server would agree. The server remained authoritative; if there was a conflict, the server won. But the *feeling* of responsiveness was transformed.

"I am sad to see the elegant client-as-terminal setup go away," Carmack wrote, "but I am practical above idealistic."

The result allowed players on terrible connections to compete with players on fast ones. It was the technical foundation that made internet gaming possible for ordinary people.

And it created a community that would outlast the game itself.

The Continuous Demand Engine

Players formed clans. They organized tournaments. They gathered for LAN parties: hauling their computers to someone's basement or garage, stringing ethernet cables across the floor, fragging each other until dawn. A competitive culture emerged, with hierarchies and rivalries and legends.

In 1997, id Software held a tournament called Red Annihilation. The prize was John Carmack's personal Ferrari 328 GTS. A player named Dennis "Thresh" Fong won it – and would later be recognized by the Guinness World Records as the first professional videogamer in history. Competitive gaming had become something a person could build a life around. Esports was born.

These players *always* wanted better performance. A faster frame rate meant a competitive edge. A higher-resolution texture meant spotting enemies sooner. The demand Doom had created became permanent, self-sustaining, insatiable.

Every year, new games pushed hardware harder. Quake II. Unreal. Half-Life. Counter-Strike. Each generation demanded more parallel processing power, more dedicated graphics capability, more of exactly what GPU manufacturers were building.

The virtuous cycle locked in: games demanded better hardware, hardware enabled better games, better games demanded even better hardware.

The Engine Business

Carmack had always shared his code.

The source for Doom was released. The source for Quake would be released. But before that happened, id Software discovered a new business model.

Other game companies wanted to use the Quake engine. Rather than spend years building their own 3D technology, they could license Carmack's work and focus on their own game design.

Raven Software. Ritual Entertainment. Hipnotic Interactive.

And Valve.

That last name would matter most. A small studio run by former Microsoft employees licensed the Quake engine to build their debut game. They modified it extensively, adding scripted sequences and narrative sophistication that id had never prioritized.

The game was called Half-Life. It shipped in 1998 and redefined what first-person games could be.

Half-Life ran on a modified Quake engine. And that engine ran best on dedicated graphics hardware. And the demand continued to compound.

Counter-Strike, the multiplayer mod built on Half-Life's foundation, would become the most popular competitive shooter in history. It was Quake's grandchild, inheriting the demand engine Carmack had created, feeding the graphics card market he had helped establish.

The Seeds

By the late 1990s, Carmack had accomplished what he set out to do.

True 3D was possible on consumer hardware. The market for graphics acceleration was established. An entire industry existed to feed the hunger he had created.

His role in this story was essentially complete.

He would continue to innovate, of course. Quake II. Quake III Arena. Doom 3. Years later, he would join Oculus VR, bringing his expertise to virtual reality – finally building toward the metaverse vision he and Abrash had discussed in 1995.

And eventually, he would turn his attention to artificial general intelligence, "the ultimate rendering problem," he might say, building minds instead of worlds.

But the foundational work was done. The demand existed. The market existed. The computational architecture, parallel processing for graphics, was locked into silicon and shipping in millions of units every year.

Carmack had made demons run faster. And in doing so, he had funded the infrastructure for something he never imagined.

. . .

In a small office in Sunnyvale, three engineers had watched all of this unfold. They had founded a company in 1993, the same year Doom shipped. They had nearly gone bankrupt twice. They had bet everything on building chips fast enough to render what Carmack was demanding.

The company was called Nvidia. Its CEO was an immigrant from Taiwan who had spent his childhood cleaning toilets in rural Kentucky. He would wake up afraid every morning for the next three decades.

His name was Jensen Huang, and he was about to become the most important semiconductor executive of his generation.

His story comes next.

Meanwhile, in Toronto, a British professor was watching none of this. Geoffrey Hinton didn't play video games. He was working on neural networks, a technology most of the field had abandoned, twice declared dead, surviving only through his stubborn refusal to accept consensus.

What Hinton needed was hardware capable of performing millions of simple mathematical operations in parallel. The same hardware that rendered Quake's monsters. The same hardware that Nvidia was building to satisfy Carmack's demands.

He didn't know this yet. Neither did Carmack. Neither did Jensen Huang.

They were all building the same cathedral. None of them knew it.

In October 2012, their work would converge. A moment that changed everything.

Hinton's story comes later.

3

THE SURVIVOR

Thirty Days from Going Out of Business

Jensen Huang wakes up afraid.

Not anxious. Not worried. *Afraid.* The bone-deep certainty that everything he's built could collapse tomorrow. That competitors are gaining while he sleeps. That the next product cycle could be the one that kills the company.

He has felt this way for thirty-three years.

"The phrase 'thirty days from going out of business,'" he told Joe Rogan in 2025, "I've used for thirty-three years." Rogan, incredulous, interrupted: "You still feel that?"

"Oh, yeah. Oh, yeah. Every morning."

Nvidia is now a four-trillion-dollar company. Jensen's net worth exceeds $100 billion. His leather jacket is iconic. His keynotes draw thousands. The infrastructure he built powers artificial intelligence everywhere.

And still, every morning: thirty days from going out of business.

"I have a greater drive from not wanting to fail than the drive of wanting to succeed," he admits. "I'm not ambitious, for example. I just want to stay alive, Joe. I want the company to thrive."

Most success coaches would call this pathological. The self-help shelves are stacked with books about visualization, abundance mindsets, manifesting your best self. Fear of failure is supposedly the enemy of achievement.

Jensen disagrees. The fear is fuel.

Where does that come from? Where does someone learn to convert fear into fuel, anxiety into action, vulnerability into strength?

The answer starts in a town of six hundred people in the poorest county in America, with a nine-year-old boy cleaning toilets for a hundred teenage strangers.

The Letter

Jensen Huang was born in Tainan, Taiwan, in 1963.

His father was a chemical and instrumentation engineer, brilliant at factory design, at the precise mechanics of industrial systems. When Jensen was young, his father took a job starting an oil refinery in Bangkok. The family moved to Thailand.

Jensen remembers Bangkok as stability. His parents working hard. A normal childhood unfolding. Then, in 1973 or 1974 (he's not precise about the year), tanks appeared in the streets.

Thailand had coups regularly. The military would rise up, seize control, then life would continue. But this time, with soldiers on every corner and the future suddenly uncertain, Jensen's parents made a decision.

The children had to leave.

"My parents thought, you know, probably isn't safe for the kids to be here," Jensen recalls. "And so they contacted my uncle."

The uncle lived in Tacoma, Washington. Jensen had never met him. Neither had his older brother Jeff.

Their parents put them on a plane anyway.

Jensen was about to turn nine. Jeff was almost eleven. Two boys, alone, flying across the Pacific to stay with a relative they'd never seen, in a country they'd never visited, speaking English their mother had taught them in preparation – enough to manage, though far from fluent.

Their uncle searched for a school that would accept foreign students and that their parents could afford. The options were limited. Their parents had almost no money.

He found Oneida Baptist Institute.

The First Night

The school was in eastern Kentucky, in Clay County, which Jensen describes as "the epicenter of the opioid crisis today, coal country." The poorest county in America when he arrived. The poorest county in America now.

"No traffic light," Jensen recalls. The town had about six hundred residents. "It's kind of an amazing feat actually. The ability to hold your population for, when it's six hundred people, it's quite a magic."

The school's mission was open admission. Anyone could come. Troubled students. Kids from broken homes. International students whose parents could afford nothing else.

Jensen's uncle drove them to Kentucky. He dropped them off.

And then Jensen was alone.

"My first thought was, 'Gosh, there are a lot of cigarette butts on the ground.'"

A hundred percent of the kids smoked. Jensen was the youngest student. Most were teenagers, junior high and high school. Tough kids from Appalachia and beyond, sent to Oneida because they had nowhere else to go.

"I was the youngest kid in school. My roommate was seventeen years old."

He pauses when he tells this story. Seventeen. Just turned seventeen. Jacked, muscular, hardened.

"That night, we got... And the second thing I noticed when you walk into your dorm room is there are no drawers and no closet doors, just like a prison. And there are no locks so that people could check up on you."

Jensen walked into his room. His roommate was getting ready for bed.

"And he had all this tape all over his body."

The boy had been in a knife fight. Stab wounds across his torso, wrapped in bandages. Fresh. The other kids had been hurt worse.

"He was my roommate, the toughest kid in school, and I was the youngest kid in school."

The Chores

Every kid at Oneida Baptist had chores. The school operated like a working institution: part boarding school, part agricultural compound, part labor exchange. Students earned their keep.

Jensen's brother Jeff worked the tobacco farm. The school grew tobacco to raise money. Child labor, essentially, with the children as the labor force.

Jensen's job was simpler. He cleaned the dorm.

"And so I was nine years old, I was cleaning toilets, and for a dorm of a hundred boys, I cleaned more bathrooms than anybody."

He pauses, thirty years later, and adds: "I just wish that everybody was a little bit more careful."

A hundred teenage boys using bathrooms that he had to scrub. Every day. The youngest cleaning up after the oldest.

The town kids all carried knives. Everyone smoked. Everyone had a Zippo lighter. Jensen bought a pack of cigarettes because everybody else did. He learned to blow smoke rings, to exhale through his nose.

"And then I did it for a couple of weeks, I guess, and I just rather have – I had a quarter, you know, I had a quarter a month or something like that, I just rather buy popsicles."

He chose popsicles over cigarettes. The better path.

The Tape Deck

The detail that haunts Jensen isn't the knife fights or the toilets or the isolation. It's simpler than that.

"We didn't have enough money to make international phone calls every week."

His parents couldn't afford to talk to their children. So they gave Jensen and Jeff a tape recorder before they left: an Aiwa tape deck and a single cassette.

Every month, the two brothers would sit in front of that machine and talk. They'd record what they'd done. What they'd seen. What their lives were like in this strange new country. Then they'd mail the tape to Thailand.

Their parents would listen. Then they'd record over it, their own voices replacing their children's, and send it back.

Back and forth, every month, for two years. The same tape, overwritten again and again. The only connection.

"Could you imagine if for two years, if that tape still existed, of these two kids just describing their first experience with the United States?"

Jensen remembers what he told them. He joined the swim team. His roommate was really buff, so Jensen started working out: a hundred push-ups every night, a hundred sit-ups. He was getting fit.

He joined teams strategically. If you joined the soccer team or the swim team, they'd take you to meets. And after meets, they'd take you to restaurants.

"And that nice restaurant was McDonald's."

He recorded it for his parents, voice full of wonder: "Mom and Dad, we went to the most amazing restaurant today. This whole place is lit up. It's like the future. And the food comes in a box, and the food is incredible."

A Taiwanese boy, living in the poorest county in America, marveling at a Big Mac like it was fine dining.

The Suffering Gene

Two years. That's how long Jensen lived at Oneida Baptist Institute.

Two years of cleaning toilets. Two years of living with kids twice his age, some fresh from knife fights. Two years of mailing tapes across the Pacific and waiting for them to come back, his parents' voices replacing his own.

Then, finally, his parents made it to America. The family reunited in Tacoma. Jensen was eleven now. His parents had left everything behind in Thailand: their jobs, their home, their community. They arrived with suitcases and whatever money they had in their pockets.

His father found a job in the newspaper classifieds. He called numbers, interviewed, got hired as a consulting engineer helping companies build oil refineries, paper mills, fabrication facilities. The same work he'd done in Thailand, now in Oregon.

His mother worked as a maid.

Jensen tells another story, one that captures how hard those early years were. The family rented an apartment. They rented furniture, everything, down to the coffee table. Particle board, cheap, the kind of thing you'd throw away in a few years.

Jensen and his brother were playing around, messing with each other the way kids do. They bumped into the coffee table.

It collapsed.

"I just still remember the look on my mom's face because they didn't have any money, and she didn't know how she was going to pay it back."

A broken coffee table. Worthless furniture, rented because they couldn't afford to buy. And his mother's face: the fear of what this damage would cost them, on top of everything else.

"That kind of tells you how hard it was for them to come here, but they left everything behind, and all they had was their suitcase, and the money they had in their pocket when they came to the United States."

Jensen went to Oregon State University, not MIT, not Stanford. He married his lab partner. He worked at AMD, then LSI Logic. He learned chips. He learned business.

And somewhere along the way, the fear became fuel. The anxiety became action. The suffering became what Jensen now calls "the suffering gene."

"I was born with the work gene, the suffering gene," he has said.

The boy who cleaned toilets at Oneida Baptist Institute learned something that billions of dollars in market capitalization would never undo: survival requires believing you might not survive.

The Denny's

In January 1993, Jensen Huang sat at a Denny's in San Jose with two friends: Chris Malachowsky and Curtis Priem. All three were chip engineers. They had an idea.

Every great company needs an origin myth. Apple had its garage. Facebook had its dorm room. Amazon had a modest house in Bellevue. The settings matter less than the symbolism: a small group of believers, scrappy and underfunded, dreaming of something larger than themselves.

Nvidia has Denny's.

The restaurant remains, still serving Grand Slam breakfasts to customers who have no idea they're sitting where a trillion-dollar company was conceived.

"We had a mission statement for a company that has no chance of success," Jensen admits. "But I didn't know that in 1993. It just sounded like a good idea."

The idea was this: build a new kind of computer. Not a faster version of existing computers, but something fundamentally different. A machine optimized for processing in parallel, capable of solving problems ordinary computers couldn't solve.

"Chris and I were married," Jensen recalls. "We have kids. I already had Spencer and Madison. They were probably two years old."

Three men with families, with mortgages, with every reason to play it safe, sketching impossible dreams on napkins.

The same year, 1993, a programmer in Texas released Doom. The demand Carmack created was exactly what Nvidia would try to satisfy.

They named the company Nvidia, from the Latin "invidia," meaning envy. They wanted competitors to envy what they built.

Jensen, at thirty, became CEO. He has never stopped.

The Arcades

Here's the detail that transforms the Nvidia origin story from Silicon Valley cliché to something stranger and more human.

Curtis Priem's townhouse became their first office. Every morning, Jensen and Chris would kiss their wives goodbye, tell them they were "going to work," and drive to Curtis's place.

"We told our family we were going to work," Jensen admits. "But it was just the three of us who was going to know."

Know what, exactly?

That there was nothing to do.

When you're building a product for a market that doesn't exist, there are no customers to call. No orders to fulfill. No urgent emails demanding faster delivery. The phone doesn't ring because nobody knows they need what you're building.

"We had really nothing to do," Jensen says.

"And so after lunch, we would always have a great lunch. After lunch, we would go to the arcades."

Every afternoon, the future architects of the AI revolution would leave Curtis's townhouse and walk to a nearby arcade. Not to relax. Not to escape. To research.

"We would play the Sega Virtual Fighter and Daytona and all those games," Jensen explains, "and analyze how they're doing it, trying to figure out how they were doing that."

The technology inside those arcade cabinets was derived from flight simulators: million-dollar systems that trained military pilots and NASA astronauts. Sega's genius was realizing you could take that technology and use it to let teenagers punch each other in virtual karate matches.

"They took the guts out of a flight simulator and put it into an arcade machine," Jensen explains.

This was the crucial insight. Graphics processing was a supercomputer problem, but it didn't have to stay that way. The same techniques that cost millions for military applications could eventually be democratized, made cheap enough that regular people could have them in their homes.

And what would those regular people do with that power?

Play games.

Sega became their first partner. "In exchange for them developing their games for our computers in the PC," Jensen explains, "we would build a chip for their game console."

"And that was kind of the beginning of Nvidia getting started. And we thought we were on our way."

He pauses.

"And about a couple of years into it, we discovered our first technology didn't work."

Three Bets, All Wrong

To understand what happened next, you have to understand how badly Nvidia got it wrong.

In 1995, after two years of work and millions of dollars in venture capital, they shipped their first product to market. The chip was called NV1. It represented everything they had: their technology, their strategic partnership with Sega, the crystallized conviction of three engineers who had staked their careers on a vision of what computing could become.

All three of their major technology bets were wrong.

Not partially wrong. Not "ahead of their time." Wrong in a way that would nearly kill the company, wrong in a way that Jensen still uses as a teaching moment thirty years later.

"Three major technology choices," Jensen recalls. "All three choices were wrong. Okay, so this is how incredibly smart we were."

First: they bet on forward texture mapping instead of inverse texture mapping. Everyone else chose inverse.

Second: they bet on curved surfaces instead of triangles. The industry had converged on triangles as the fundamental polygon. Three points. One plane. Mathematically simple. Nvidia chose quadratic surfaces, curved patches that could represent smooth shapes more naturally. The demos were beautiful. Working with them was, in the words of one contemporary account, "horrendous."

Third: no Z-buffer. Applications had to sort polygons themselves before sending them to the chip.

Then Microsoft announced DirectX specifications based on triangle polygon rendering. Everything Nvidia had bet against became the industry standard.

"Everybody had chosen the right technology approach," Jensen says, "and we chose the wrong one. And so we were the first company to start. We found ourselves essentially dead last with the wrong answer."

By mid-1996, Nvidia was dying.

The Flight to Tokyo

The flight from San Francisco to Tokyo is eleven hours.

Jensen Huang had plenty of time to rehearse what he was going to say. He was thirty-three years old, still breaking out with acne, "super skinny" by his own description – a Chinese kid from Oregon State who had bet everything on a company that had just shipped the wrong product to a market that didn't want it.

He was flying to Japan to tell the CEO of Sega that everything Nvidia had promised was worthless.

The conversation he was about to have violated every rule of business negotiation. You don't admit weakness to a partner. You don't confess failure while you're still under contract. You don't go to someone who has already given you millions of dollars and tell them you can't deliver what you promised.

And you certainly don't, after admitting all of that, ask them for *more* money.

But Jensen had no other options. The math was simple: if Sega canceled the contract, Nvidia would "vaporize instantly." Those were his words. Not "struggle" or "face challenges" or "need to restructure." Vaporize. The company would cease to exist overnight.

So he sat on that plane, rehearsing the most honest pitch of his life.

The Confession

Shoichiro Irimajiri was not the kind of executive you typically found in the video game industry.

He had been the CEO of Honda USA, responsible for building Honda's American manufacturing presence at a time when Japanese companies were transforming the global auto industry. He was an engineer by training, a builder by instinct, and an elder statesman by the time Sega brought him in.

Jensen walked into that meeting carrying the weight of his company's survival.

"Listen, I've got some bad news for you."

What followed was one of the most candid confessions in business history. Jensen told Irimajiri that the technology Nvidia had promised didn't work. That Sega shouldn't finish the contract because they'd be wasting money on something broken. That Jensen recommended Sega find another partner to build their game console.

He was actively telling Sega to abandon him.

And then came the ask that defied all logic: even though he was releasing Sega from the contract, Jensen still needed the five million dollars they'd promised – because without it, Nvidia would vaporize overnight.

Irimajiri listened as Jensen walked through the situation with complete transparency. The older executive understood immediately. With dozens of competitors doing it right, what were the chances that Nvidia would succeed even with the investment?

Jensen pauses when telling this story. He knows the answer.

"Zero percent. You do the math, it's zero percent."

Irimajiri went away to think about it for a couple of days.

"He came back and said, 'We'll do it.'"

Why?

"What he decided was Jensen was a young man he liked. That's it."

Joe Rogan, hearing this story decades later: "Boy, the world owes that guy."

"No doubt. Crazy."

If Sega had kept that investment, if they had held their Nvidia shares instead of selling after the IPO, it would be worth approximately a trillion dollars today. They sold for fifteen million dollars – a three-times return on a bet everyone said was irrational. They thought they'd gotten lucky.

They had no idea.

The Textbook

With Sega's five million dollars keeping Nvidia barely alive, the team had one shot at redemption.

But there was a problem. Nobody at Nvidia knew how to build chips the right way. They had spent two years building expertise in quadratic texture mapping. Their entire engineering organization was trained on the wrong technology.

Someone inside the company voiced what everyone was thinking: "But Jensen, we've never built it this way before. We've never built it the right way before. We only know how to build it the wrong way."

Jensen had an answer.

"I said, 'Okay. How hard can it be? You've got all these thirty companies, fifty companies doing it. How hard can it be?'"

The question was rhetorical, but the answer was concrete. There was a textbook.

Silicon Graphics, the company that had invented modern computer graphics for high-end workstations, had literally written the book on how to build 3D graphics systems. Their OpenGL specification and accompanying documentation laid out the architecture that had become industry standard.

Jensen went to a store with $200 in his pocket. He bought three textbooks, $60 each, the only copies they had. He brought them back to the office and gave one to each of the company's architects.

"Read that," he told them, "and let's go save the company."

The architects read the textbooks. They learned from Silicon Graphics. And then, here's what made Nvidia special even then: they reimplemented everything in a way that had never been done before. They reimagined 3D graphics, learning from previous known arts but implementing it fundamentally differently.

Where Silicon Graphics ran their geometry engine as software on processors, Nvidia stripped out all the generality and hard-coded the essential operations directly into silicon. They sacrificed flexibility for speed.

"And our one little chip was generating images as fast as a million-dollar image generator. We took a million-dollar thing and we put it into the graphics card that you now put into your gaming PC."

The Emulator

Building a new chip normally required multiple iterations. Design it, send it to a foundry, wait for manufacturing, test it, find the bugs, redesign, send it back.

Nvidia didn't have time for multiple cycles. They didn't have money for two. They barely had resources for one.

"We did the math. There was no chance."

Then Jensen heard about a company called Ikos. They had built an emulator, a machine that could simulate your chip design before you manufactured it. You could run real software on a simulated version of your hardware and find bugs before you ever touched silicon.

Jensen called them.

"Hey, listen, I heard about this machine. I'd like to buy one."

"Oh, that's terrific. But we're out of business."

"What? You're out of business?"

"Yeah, we had no customers."

Jensen processed this. "Wait, hang on a second. So you never made the machine?"

"No, no, no, we made the machine. We have one in inventory if you want it. But we're out of business."

So Jensen bought an emulator from a company with no other customers, using half of the money Nvidia had left in the bank. About five hundred thousand dollars, half their runway, for a machine that might let them skip the traditional testing cycles.

If it didn't work, they'd die six months sooner.

After Jensen bought the machine, Ikos went out of business. If he had called one day later, the machine wouldn't have existed.

The Gamble

With the emulator, Nvidia loaded their new chip design and ran software against it. They found bugs. They fixed them. They found more bugs. They kept testing until they had convinced themselves, truly convinced themselves, that the chip would work on first silicon.

No one had ever done this before. Every chip company in the world expected multiple revision cycles.

Jensen called TSMC, Taiwan Semiconductor Manufacturing Company, the foundry that would fabricate their chips. TSMC was still small then.

"Listen, we're gonna tape out a chip to you, and I'd like to go directly to production because I know it works."

TSMC's response: "Nobody has ever done that before. Nobody has ever taped out a chip that worked the first time, and nobody starts production without looking at it."

Jensen knew that. He also knew that if he didn't start production, he'd be out of business anyway.

Morris Chang, the founder of TSMC, decided to support the small company. He flew to the United States and asked Jensen questions, questions clearly trying to determine whether Nvidia could actually pay their bills.

Jensen didn't quite lie. But he didn't tell the whole truth either.

TSMC took the risk. Jensen took the risk. Everyone held their breath.

The Prefetch

Years later, interviewers would ask Jensen about betting the company.

They'd frame it as courage: the willingness to push all your chips in, to go all-in on conviction.

Jensen corrects them.

"When you push your chips in, I know it's going to work," he told the Acquired podcast. "Notice, we assumed that we taped out a perfect chip. The reason why we taped out a perfect chip is because we emulated the whole chip before we taped it out."

This wasn't gambling. This was engineering.

"We developed the entire software stack. We ran QA on all the drivers and all the software. We ran all the games we had. We ran every VGA application we had."

The insight that emerged from this desperate moment became a permanent philosophy:

"When you bet the farm, you're saying, 'I'm going to take everything in the future, all the risky things, and I'm going to pull them in advance.' And that is probably the lesson. And to this day, everything that we can prefetch, everything in the future that we can simulate today, we prefetch it."

Betting the company, in Jensen's view, isn't about risk tolerance. It's about eliminating risk *before* you bet. Simulate the future. Find the failures. Fix them. Then push your chips in – not because you're brave, but because you already know.

"So every time we see you make a bet the company move," the interviewer observed, "you've already simulated it. You know."

"Yeah, yeah, yeah."

RIVA 128

The chip shipped in August 1997. It was called RIVA 128.

It worked.

The first silicon came back functional. No revision cycles needed. No additional months of debugging. Just a working chip, ready for market.

Where NV1 had been different for the sake of being different, RIVA 128 was right in exactly the ways that mattered. Triangles instead of curved surfaces. Full compatibility with DirectX and OpenGL. Everything the industry had standardized on, implemented with ruthless efficiency.

The chip was fast. It ran Quake, the game that had consumed John Carmack for three years, at framerates that made gamers weep with joy.

"We became the fastest growing technology company in history to go from zero to one billion dollars."

The methodology mattered as much as the product. The emulator-based development process, born from desperation, became Nvidia's permanent approach.

"That methodology that we developed to save the company is used throughout the world today. We changed the whole world's methodology of designing chips, the whole world's rhythm of designing chips. We changed everything."

The Word

In early 1999, Nvidia held a contest called "Name That Chip."

They were preparing to launch their most powerful graphics chip yet, the successor to RIVA TNT2. It would feature something no consumer card had ever included: a dedicated hardware engine for transform and lighting calculations.

Over twelve thousand entries arrived. The winning name was "GeForce," a portmanteau of "Geometry Force."

But the product name was almost beside the point. What mattered was the category name.

Brian Burke, Nvidia's senior PR manager, explained the thinking: they needed a term that would distinguish their chip from ordinary "graphics accelerators." Something that conveyed the leap in capability.

On August 31, 1999, Nvidia announced the GeForce 256. The press release declared it "the world's first Graphics Processing Unit, or GPU."

At the time, tech journalists thought it was pompous. How could this relative newcomer have the nerve to think it could change the language of PC hardware?

Pompous or not, the term stuck. Before GeForce, there were graphics cards. After GeForce, there were GPUs.

This is the power of naming. Whoever names the thing owns the thing. Intel had the CPU. Apple had the smartphone. Now Nvidia had the GPU.

Twenty-five years later, the term describes not just gaming hardware but the processors that train neural networks, render Hollywood films, simulate climate models, and power the AI revolution. The very language of the industry encodes Nvidia's ownership.

All because someone in marketing, preparing for a chip launch in 1999, decided that "graphics accelerator" wasn't impressive enough.

The Collapse

While Nvidia was rising, 3dfx was dying.

3dfx had invented the category. Their Voodoo Graphics cards had made consumer 3D gaming possible. They controlled 85% of the 3D graphics market in 1997. They were the industry.

They lost anyway.

The reasons cascade like dominoes. 3dfx clung to their proprietary Glide API while the industry standardized on DirectX. They couldn't ship a competitive 2D/3D combo card. They made a disastrous strategic decision, acquiring STB Systems, a graphics card manufacturer, for $141 million.

The consequences were immediate. Diamond Multimedia, Creative Labs, and other major board manufacturers, the companies that had built 3dfx's brand, defected to Nvidia. Why would they buy chips from a competitor who was now trying to steal their customers?

3dfx alienated their entire ecosystem to capture a little more margin. It was suicide by spreadsheet.

On December 15, 2000, 3dfx announced its end. Facing bankruptcy, the company sold its brands and assets to Nvidia. The technology that had launched consumer 3D graphics was absorbed by the company that had almost died in 1996.

Jensen's paranoia had been vindicated. He never felt safe. He never stopped running. His competitor, the one who had seemed unbeatable, had stopped running.

In competition, as in evolution, survival is not about being strongest. It's about not being the one who dies.

Zero Billion Dollar Markets

After 3dfx's collapse, Jensen began articulating a philosophy that would define Nvidia's next two decades.

"We prefer to position ourselves in a way that serves a need that usually hasn't emerged," he explains. "Zero billion dollar markets. It's our way of saying there's no market yet, but we believe there will be one."

When Nvidia entered automotive, building computers for cars that didn't exist yet, industry executives told Jensen it was foolish. "One of the CTOs told me, 'You know what, cars cannot tolerate the blue screen of death.'"

Jensen's response: "I don't think anybody can tolerate that, but it doesn't change the fact that someday every car will be a software-defined car."

Fifteen years later, he was right.

"And so oftentimes there's non-consumption, and we like to navigate our company there. And by doing that, by the time that the market emerges, it's very likely there aren't that many competitors shaped that way."

It was a strategy born from near-death. When you've survived bankruptcy twice, you learn to see around corners – to build for markets that don't exist yet, because by the time everyone else sees the opportunity, it's already too late.

The Lesson

"I learned how to develop strategies," Jensen reflects on the NV1 disaster and the RIVA redemption. "When our company learned how to develop strategies, what are winning strategies? We learned how to create a market. We created the modern 3D gaming market."

Then, crucially: "That exact same skill is how we created the modern AI market. It's exactly the same. It's exactly the same skill, exactly the same blueprint."

Start with what you know is wrong. Stop doing it. Learn from whoever got it right. Reimagine it in your own way. Move faster than anyone thinks is possible. Survive.

"And we learned how to deal with crisis, how to stay calm, how to think through things systematically. We learned how to remove all waste in the company and work from first principles and doing only the things that are essential. Everything else is waste because we have no money for it. To live on fumes at all times."

That feeling of living on fumes? It never left. The company grew. The resources grew. The anxiety stayed the same.

The Vindication

By 1999, Nvidia had gone public, invented the GPU, and absorbed its greatest competitor's assets. The market for graphics processing, the market that John Carmack's demons had helped create, now belonged to Jensen Huang.

The kid who had cleaned toilets in the poorest county in America. The immigrant who had marveled at McDonald's like it was fine dining. The CEO who woke up every morning convinced his company was about to fail.

He had survived.

But survival wasn't enough. Jensen had learned something in those years of near-death and redemption. He had learned that the real victories come from building something nobody else can see yet.

The demand was there – John Carmack and id Software had seen to that. Every teenager who played Doom wanted faster graphics. Every Quake deathmatch pushed hardware to its limits. Nvidia didn't create the hunger. Carmack did. Jensen's job was to feed it.

And now, having fed it, he was about to do something that seemed insane.

He was going to bet over a billion dollars on a technology nobody wanted, for a market that didn't exist, serving customers who didn't know they needed it.

The same pattern. The same conviction.

A different cathedral.

4

THE BET

The Quake Connection

Picture this: a Thursday night at Stanford University, around the year 2000. A PhD student named Ian Buck is staring at an impossible problem.

He wants to play Quake at a resolution no monitor can display – rendering it across eight projectors simultaneously, creating an immersive panorama of Carmack's dark corridors. The hardware doesn't exist. So Buck builds it. He connects thirty-two GeForce graphics cards into a single rendering farm, linking them with cables and prayers and the kind of stubborn ingenuity that defines a certain type of computer scientist.

It works. Sort of. The frame rates are terrible, but the image, spread across multiple screens, surrounding the player in John Carmack's dark corridors, is spectacular.

But here's what makes Buck different from any other gamer pushing hardware to its limits: he starts asking the wrong question.

Instead of "How do I make Quake run faster?" he asks, "What else could these cards do?"

Buck had interned at Nvidia before pursuing his PhD. He knew what GPUs were designed for: rendering triangles, computing lighting, pushing pixels. Gaming hardware for gaming purposes. But staring at his thirty-two-card monstrosity, he saw something else. He saw thousands of tiny processors, all working simultaneously. He saw parallel processing on a scale that no CPU could match.

He saw a supercomputer hiding inside consumer hardware.

"At the time, a lot of the GPU development was driven by the need for more realism," Buck has explained, "which meant programs were being written that could run at every pixel to improve the game. These programs were tiny then – four instructions, maybe eight – but they were running on every pixel on the screen; a million pixels, sixty times per second."

A million operations per frame. Sixty frames per second. That's sixty million tiny calculations every second, just to make demons look scary.

What if you could harness that power for something other than demons?

The Bridge

Buck's PhD thesis, *Brook for GPUs: Stream Computing on Graphics Hardware*, introduced something radical: make it possible for ordinary programmers, not graphics experts, not shader specialists, to write code that ran on GPUs. Abstract away the graphics-specific details. Let scientists and researchers access those thousands of parallel processors without needing a PhD in computer graphics.

Buck's work attracted support from DARPA, the same defense research agency that had funded the early internet. It also attracted attention from Nvidia.

Jensen Huang recognized something in Buck's research. Here was a way to expand GPUs beyond gaming – to transform a successful graphics chip company into something much more valuable. In 2004, Nvidia hired Buck away from Stanford.

"Nvidia was already very supportive of my research and clearly saw the potential to better enable GPU computing on the hardware side of things," Buck has recalled. "I joined Nvidia to start the CUDA project in 2005."

The project that would cost over a billion dollars. The bet that would take a decade to pay off. The infrastructure for a revolution nobody saw coming.

It started with a guy who wanted to play Quake really, really well.

The Pattern

Here's what connects Ian Buck to John Carmack to Geoffrey Hinton, though none of them saw it at the time:

The same computational architecture that Carmack had exploited to render demons – the embarrassingly parallel processing we encountered earlier, thousands of simple operations happening simultaneously – turned out to be exactly what Buck needed for general-purpose computing.

Think back to Chapter 1. The restaurant kitchen analogy. A CPU is a single brilliant chef, preparing dishes sequentially. A GPU is a thousand line cooks, each doing one simple thing, all working at once.

GPUs were already optimized for this. Gaming had demanded it. Every frame of Quake required computing the color of hundreds of thousands of pixels, and those calculations didn't depend on each other. You could compute them all at once if you had enough processors.

Neural networks, it would turn out, worked the same way. Matrix multiplications. Weight updates. Backpropagation. Millions of simple operations, embarrassingly parallel, capable of running simultaneously on thousands of cores.

But in 2006, nobody knew this yet. The connection between gaming graphics and artificial intelligence was invisible. Jensen was betting on a capability, not a specific application.

The kid playing Doom in 1994, demanding faster frame rates, had unknowingly funded the architecture that would train ChatGPT.

The Silent Audience

By 2006, Nvidia dominated gaming graphics. They had won the war Carmack started. They were profitable, growing, secure.

Jensen made a bet that seemed insane.

He committed years of R&D investment to CUDA: Compute Unified

Device Architecture. A technology that would let scientists and researchers use GPUs for general-purpose computing.

The gaming business was working. Why bet on something else?

"When I launched CUDA," Jensen has recalled, "the audience was complete silence. No customer wanted it, nobody asked for it, nobody understood it."

Imagine that moment. You're the CEO of a publicly traded company. You've just invested years of engineering effort and hundreds of millions of dollars into a technology. You stand on stage to announce it. And the audience – your customers, your investors, the analysts who will write reports about your company – sits in bewildered silence.

"Nvidia was a public company," Jensen continued. "Our stock prices went poof. I think our valuation went down to like two billion or three billion from about twelve billion."

Jensen had crushed his own stock price by building something nobody wanted.

"I crushed it in a very bad way."

Reasoned Hope

There's a philosophy embedded in the CUDA bet that Jensen has articulated across dozens of interviews, refining the language each time.

"When you create something new like CUDA, if you build it, they might not come," he has explained. "And that's always the cynic's perspective. However, the optimist's perspective would say, 'But if you don't build it, they can't come.'"

This wasn't blind optimism. Jensen calls it "reasoned hope," a strategy grounded in first principles, not wishful thinking.

"You have to reason from your best principles," he has said. "And ideally, you're reasoning from principles of either physics or deep understanding of the industry or deep understanding of the science. You reason from first principles. And at some point, you have to believe something."

The reasoning went like this: Nvidia's GPU was going to be the highest-volume parallel processor built in the world because the gaming market was so large. If parallel computing had any future beyond graphics, and the physics suggested it did, then this architecture had the best chance of reaching the people who would find it.

"It was a strategy based on hope, but it was also reasoned hope."

The Cost of Conviction

To understand the magnitude of the CUDA bet, you have to understand what Jensen was giving up.

Every GPU Nvidia shipped now included CUDA capability: additional silicon that cost money to manufacture but generated zero revenue. The chips got more expensive. The margins got thinner. Customers who just wanted to play World of Warcraft were subsidizing technology they would never use.

"We added this thing called CUDA to the chip," Jensen explained to Joe Rogan, "but nobody paid for it, but our cost doubled."

The financial commitment was staggering. "It was ten long years," Jensen has said.

Wall Street didn't get it.

"Some investors were big Nvidia fans in the late 2000s and gave them the benefit of the doubt for the first five years of the CUDA investment," one analyst noted. "But in the mid-2010s, market demand still wasn't showing up in a big way, and it was becoming a bigger and bigger investment."

For the better part of a decade, Nvidia's market cap held relatively flat, lingering between $8.5 billion and $10 billion from 2009 to 2015. The stock was dead money. Analysts questioned why the company was wasting resources on something with no obvious market.

Bryan Catanzaro, Nvidia's vice president of applied deep learning research, joined the company in 2008, right in the middle of the CUDA bet. His sense of that time: Nvidia and Huang were the underdogs. Wall Street fully expected CUDA to be a company-ending failure.

"It made all of our products more expensive since we were selling these gamer cards while putting computing acceleration into them," Catanzaro recalled. "It took a lot of commitment to follow through. I would say it was about ten years before Wall Street really started to believe this investment was worth anything."

Ten years. A decade of waiting. A decade of faith.

The Psychology of Belief

Why did Jensen keep investing?

"If you believe in that future," he has said, "and you don't do anything about it, you're going to regret it for your life."

The language of conviction, not business strategy.

"And so I always tell the team, 'Do we believe this or not?' And if you believe it, and grounded on first principles, not random hearsay, and we believe it, we owe it to ourselves to go pursue it. If we're the right people to go do it, if it's really, really hard to do, it's worth doing, and we believe it, let's go pursue it."

Jensen's bet was not on artificial intelligence. In 2006, AI was still in the grip of its second winter. Neural networks were a backwater. Deep learning was an academic curiosity. The word "ChatGPT" would not exist for another sixteen years.

Jensen's bet was on parallel computing itself. He believed that the architecture Nvidia had built for gaming – massive parallelism, thousands of cores, embarrassingly parallel processing – was too powerful to be limited to pixels. Somewhere, someone would figure out what to do with it.

"We were building infrastructure for a future we couldn't see," he has said.

There's a psychological dimension to the CUDA bet that the business press rarely captures.

Remember Jensen's morning anxiety – the perpetual feeling of being thirty days from going out of business? The CUDA bet makes sense through that lens. Jensen wasn't optimizing for shareholder value or quarterly earnings. He was building for survival: not today's survival, but the survival that comes from having capabilities nobody else has. He was building a moat before the castle was attacked.

"Always in a state of anxiety. Every morning. Every moment."

And so he invested in CUDA through the global financial crisis. Through the stock price crater. Through the years of analyst complaints and investor defections. Through the long silence after every announcement.

The Early Adopters

The early adopters were scattered across academia.

Physicists modeling climate. Chemists simulating molecular dynamics. Astronomers processing telescope data. At Stanford, Professor Todd Martínez pioneered GPU-accelerated quantum chemistry. At universities across the world, small research groups discovered that simulations that took months on traditional computers could run in hours on CUDA-enabled GPUs.

One quantum chemistry scientist told Jensen something that captures what CUDA meant to its early users: "Jensen, because of Nvidia's work, I can do my life's work in my lifetime."

That's what Jensen calls "time travel." A GPU lets you see the future sooner. It compresses decades of computational work into years, years into months.

These researchers formed a community. Nvidia cultivated them, provided free hardware to academic labs, sponsored conferences, built libraries and tools. The GTC – GPU Technology Conference – became an annual gathering where scientists shared what they'd accomplished with parallel computing.

The revenue was negligible. The customers were academics who couldn't afford much. The applications were esoteric: protein folding, atmospheric modeling, computational fluid dynamics. Wall Street saw no path to profit.

"We democratized supercomputing," Jensen has said of this period. "CUDA is not just used for AI. CUDA is used for almost all fields of science – everything from molecular dynamics to imaging, CT reconstruction to seismic processing to weather simulations, quantum chemistry."

He kept investing anyway.

The Proprietary Ecosystem

CUDA was free. Anyone could download it. Any programmer could learn it.

But CUDA only ran on Nvidia hardware.

This was Jensen's masterstroke, the strategic genius hiding inside the charitable gesture. By making CUDA free, Nvidia removed the barrier to adoption. Scientists and researchers could experiment without budget approval. Graduate students could learn on their laptops. The developer community could grow organically.

But every line of CUDA code written was a line of code that only ran on Nvidia GPUs. Every library built for CUDA was a library that didn't work on AMD or Intel. Every scientist who invested months learning the platform was a scientist locked into Nvidia's ecosystem.

The Acquired podcast's analysis of Nvidia captures the strategy precisely: "Yes, it is free to use. But it is also a proprietary language that only works on Nvidia GPUs."

By 2025, CUDA would support more than 900 libraries and AI models. Each library corresponds to an industry using Nvidia technology. The software moat, built during the years when nobody cared, became the moat that competitors couldn't cross.

The bet wasn't just on parallel computing. It was on creating an ecosystem so deep and so comprehensive that switching costs would become prohibitive. By the time the killer app arrived – by the time neural networks needed exactly what CUDA provided – the infrastructure was already there, and everyone who mattered was already trained on it.

The Waiting

From 2006 to 2012, Jensen waited.

The scientists used CUDA. The publications accumulated. The research community grew. But the breakthrough application, the killer app that would justify a decade of investment, hadn't arrived.

And then it did.

In Toronto, a British professor named Geoffrey Hinton had been working on neural networks for forty years. Two AI winters had frozen the field around him. Most of his colleagues had moved on to more fashionable approaches. But Hinton kept working, and he had students – bright young people who believed in the mathematics of learning.

In 2012, two of those students, Alex Krizhevsky and Ilya Sutskever, built something that changed everything. They called it AlexNet. It was a neural network trained on millions of images, capable of recognizing objects with accuracy that crushed every previous approach.

The training had taken days instead of months. The cost had been minimal. The hardware?

Two $500 Nvidia graphics cards. Consumer gaming hardware. The same silicon that Ian Buck had once chained together to play Quake across eight projectors.

They had done what everyone thought was impossible, on hardware anyone could buy at Best Buy. And the only reason they could do it was because Jensen Huang, six years earlier, had made a bet nobody understood.

The neural network revolution didn't need supercomputers. It needed CUDA.

The Recognition

Jensen was paying attention.

Nvidia had been working on computer vision at the time, struggling to make CUDA work for traditional algorithms. Jensen's team was frustrated by the limitations of conventional approaches. Then AlexNet appeared – a completely different method that was a perfect fit for what they'd built.

Jensen stepped back to reason through the implications. A new algorithm had just leapfrogged thirty years of computer vision research. Why? Was it scalable? If scalable, what other problems could it solve?

The answer was profound: they had discovered a universal function approximator. The dimensionality could be as high as needed. Each layer trained on top of the previous one, meaning the networks could go arbitrarily deep.

A computer you could teach rather than program. And it ran on the hardware Jensen had been building in silence for six years.

Finding the Researchers

Jensen didn't wait for the researchers to come to him. He went to them.

Jensen went directly to the researchers – Yann LeCun, Andrew Ng, Geoffrey Hinton – asking how Nvidia could help advance their work. He attended every AI conference, personally meeting researchers whose work wouldn't generate meaningful revenue for years. Most CEOs would delegate this. Jensen showed up. That's where he first met Ilya Sutskever.

He met researchers working on ideas that seemed impractical. Ian Goodfellow, a PhD student who would later work at Google Brain and Apple, had recently invented GANs – generative adversarial networks, a technique that could teach machines to create images from scratch. When Jensen first saw it, the GAN produced thirty-two by thirty-two pixel images – just blurry impressions of cats. A toy.

But Jensen saw the trajectory. Deep learning trained layer by layer. Datasets could grow larger. Models could grow larger. If you scaled it up, it would get better. The question wasn't whether the technology was ready – it was how far it could go.

The First DGX

By 2015, Jensen saw where this was heading. He started building something new: the DGX, a complete AI supercomputer in a box. The first version was used internally – a seventy-pound machine with thirty-five thousand parts drawing ten thousand amps.

When OpenAI was founded in 2015, Jensen knew what they needed. He personally delivered the first DGX to the fledgling organization. The CEO of Nvidia, hand-delivering the first AI supercomputer to the startup that would eventually create ChatGPT.

Nobody knew then what OpenAI would become. Nobody knew that this hardware would help train models that would change how humanity interacts with computers. But Jensen believed. The technology wasn't useful in its current state. In a few iterations, it could change everything.

The Vindication

The CUDA bet changed the world. The stock price went higher, the market cap exploded, the valuation soared – but the real vindication was watching the technology proliferate.

The vindication came suddenly after years of waiting. AlexNet demonstrated that GPUs could train neural networks orders of magnitude faster than CPUs. Researchers everywhere took notice. Suddenly, everyone needed what Nvidia had been building in silence.

Jensen saw it cropping up everywhere. Machine learning, AI, neural networks – these had been around for decades. What made now the moment? The answer was parallel processing. The architecture Nvidia had built for gaming, the architecture Carmack had demanded, was exactly what deep learning required.

The realization crystallized: deep neural networks and backpropagation fit the GPU architecture perfectly. The solution could scale. It could solve many problems.

The Cathedral Connection

Here is the hidden structure that nobody saw until it was too late:

John Carmack, in 1993, wanted to make demons run faster. His pursuit of frame rates created the demand for parallel processing hardware.

Jensen Huang, in 2006, wanted to expand beyond gaming. His CUDA bet created the platform for general-purpose GPU computing.

Geoffrey Hinton, in 2012, wanted to prove neural networks could work. His students' discovery revealed that everything Carmack demanded and Jensen built was exactly what AI required.

Ian Buck, playing Quake at Stanford, was the hinge. He saw that gaming hardware could become computing hardware. He built the bridge between entertainment and science. He connected Carmack's demons to Hinton's neurons.

None of them planned this. Carmack wasn't thinking about AI. Jensen wasn't betting on neural networks. Hinton wasn't designing for GPUs. Buck started with a gaming rig and saw a supercomputer.

The kid playing Doom in 1994, demanding faster graphics. The scientist running molecular simulations in 2008, grateful for CUDA acceleration. The AI researcher in 2024, training models on clusters of Nvidia H100s. They're all part of the same story.

A story about how tools for play became tools for thought.

The Blueprint

"That exact same skill is how we created the modern AI market," Jensen has said, connecting the CUDA bet to the lessons learned from the NV1 disaster. "It's exactly the same. It's exactly the same skill, exactly the same blueprint."

Start with a conviction grounded in first principles. Build the infrastructure before the demand exists. Keep investing through the silence. Wait for the world to catch up.

The NV1 disaster taught Jensen to pivot brutally when he was wrong. The CUDA bet taught him to hold firm when he was right. Both required the same underlying skill: the ability to distinguish between a wrong strategy and a strategy that hasn't worked *yet*.

"The question of why did we stay so committed for so long," Jensen has reflected, "the answer is actually the opposite. There was no reason to not be committed because we believed it."

The CUDA bet took ten years to pay off. From the perspective of quarterly earnings, it was a disaster. From the perspective of strategic positioning, it was the most important decision in Nvidia's history.

While Jensen was betting on CUDA, a British professor in Toronto was betting on something even more uncertain.

Geoffrey Hinton had been working on neural networks since the 1970s. Two AI winters had frozen the field around him. Most of his colleagues had abandoned ship. By 2006, advocating for neural networks was professional suicide.

Hinton didn't care. He kept refining the same fundamental mathematics, training small networks on limited datasets, proving concepts that would scale once the infrastructure existed.

What Hinton needed was hardware capable of performing millions of simple mathematical operations in parallel. The same hardware that rendered Quake's monsters. The same hardware that Jensen was making programmable with CUDA.

He didn't know about CUDA yet. Jensen didn't know about Hinton yet.

The supply was searching for demand. The demand was waiting for supply.

Three threads. One cathedral.

Hinton's story comes next.

5

THE HERETIC

The Boole Problem

Picture a family tree that reads like a history of human thought itself.

Your great-great-grandfather invented the mathematics that makes computers possible – Boolean logic, the binary yes-or-no at the heart of every digital system. His algebra of True and False, AND and OR, underlies every circuit, every line of code, every calculation that has ever run on a machine. George Boole algebraized thought itself.

Your great-great-granduncle surveyed the highest mountains on Earth. George Everest, whose name graces the peak that represents the ultimate physical achievement.

Your great-grandfather coined the word "tesseract" and spent his life trying to help people visualize the fourth dimension. Charles Howard Hinton, mathematician, mystic, author of "scientific romances" that influenced H.G. Wells. He believed our three-dimensional reality was merely a slice of a higher-dimensional truth.

Your father is a distinguished entomologist, Fellow of the Royal Society, known for his intimidating intellect. The family ethos, passed down through generations of scholars and surveyors and mathematicians, is simple, brutal, and binary, fittingly, given your ancestor.

Be an academic or be a failure.

There is no middle ground.

Geoffrey Everest Hinton was born in London in 1947, carrying his family's intellectual ambitions in his very name. The "Everest" came from Mary Everest Boole, George Boole's wife, niece of the mountain surveyor, grandmother of the tesseract visionary.

"Be an academic or be a failure," his mother told him. He heard the message. He would spend his life in academia.

But he would also spend that life on a peculiar quest: proving that his great-great-grandfather's logic wasn't enough.

The Question

From early childhood, Hinton was fascinated by a question so fundamental that it almost seems naive: How does meat think?

You can hold a brain in your hands. You can slice it, stain it, measure its electrical activity. It's tissue: neurons and axons, synaptic gaps and chemical gradients. Physical stuff. About three pounds of it, the consistency of firm tofu.

Yet somehow, inside that lump of matter, there is experience. There is memory. There is the voice in your head reading these words right now.

How?

The question drove him through a wandering academic path: physics first, then physiology, then philosophy, finally settling on experimental psychology at Cambridge. But really, he was circling the same obsession from different angles.

"I was trying to come up with ideas about how the brain actually learned," Hinton explains, decades later. "There's some things we know about that. It learns by changing the strengths of connections between brain cells."

The insight seems simple now. The brain isn't a fixed machine running a program. It's a plastic network. When you learn something new, you don't add new hardware; you adjust the connections that already exist.

Think of it this way: every neuron in your brain "can sometimes go ping," as Hinton puts it. That's all it does. A single neuron has one job. It receives signals from other neurons, and if enough of them are pinging in the right pattern, it pings too.

"And it has to decide when to go ping."

How does it decide? By tallying votes. Each incoming connection has a strength, a weight. When other neurons fire, they cast votes for or against this neuron's firing. If the weighted sum exceeds a threshold, the neuron fires. If not, it stays quiet.

Learning, then, is adjusting those weights. Make this connection stronger. Make that one weaker. The more you're exposed to a pattern, the more the network adjusts to recognize it.

"It's very like political coalitions," Hinton says. "There'll be groups of neurons that go ping together. And the neurons in that group will all be telling each other go ping. And then there might be a different coalition, and they'll be telling other neurons don't go ping."

Your concept of "spoon" isn't stored in any single neuron. It's a coalition, a particular pattern of neurons that tend to fire together when you see a spoon, hear the word "spoon," imagine using a spoon. The concept is distributed across the network.

The same neurons might also fire for "fork." The coalitions overlap. That's why similar things feel similar; they share neural real estate.

This is a radically different picture of mind than the one that dominated artificial intelligence in the 1970s. The AI establishment believed that intelligence was symbol manipulation, that the brain was essentially running logical rules, like a computer executing a program.

His great-great-grandfather's algebra, in other words.

Hinton thought they had it exactly backward.

The Two Churches

"For a long time in AI, from the 1950s onwards, there were kind of two ideas about how to do AI."

Hinton describes them like competing religions. Two different faiths about the nature of intelligence. Two different visions of what computers should become.

"One idea was that sort of core of human intelligence was reasoning, and to do reasoning you needed to use some form of logic. And so AI had to be based around logic. And in your head, you must have something like symbolic expressions that you manipulated with rules, and that's how intelligence worked."

This was the dominant church. Its high priests were giants: John McCarthy, who coined the term "artificial intelligence." Marvin Minsky, who ran the MIT AI Lab. Newell and Simon, who proposed the Physical Symbol System Hypothesis – the idea that symbol manipulation is both necessary and sufficient for general intelligence.

Under this paradigm, if you wanted a machine to understand cats, you'd write rules: a cat has four legs, whiskers, retractable claws, and says "meow." You'd encode this knowledge explicitly, logically, symbolically. The machine would reason about cats by manipulating these symbols according to formal rules.

"Things like learning or reasoning by analogy, they'd all come later once we've figured out how basic reasoning works."

This approach had prestige. It had funding. It had the imprimatur of MIT, Stanford, Carnegie Mellon. It had the backing of DARPA, which was pouring money into AI with the expectation of military applications. It had, whether anyone acknowledged it or not, the blessing of George Boole himself.

"There was a different approach," Hinton continues, "which is to say, let's model AI on the brain, because obviously the brain makes us intelligent. So simulate a network of brain cells on a computer and try and figure out how you would learn strengths of connections between brain cells so that it learned to do complicated things, like recognize objects and images or recognize speech or even do reasoning."

This was the minority church. The connectionists. The heretics.

They didn't want to program rules. They wanted to build learning machines, systems that would figure things out for themselves by adjusting their own internal connections.

"It was sort of obvious to a lot of people that the brain doesn't work by someone else giving you rules, and you just execute those rules," Hinton says. "I mean, in North Korea, they would love brains to work like that, but they don't."

Babies learn to recognize faces before they can understand rules. Toddlers master language without being taught grammar explicitly. The brain doesn't wait for instructions. It extracts patterns from experience and adjusts its own wiring.

Surely, Hinton thought, artificial intelligence should work the same way.

"I pushed that approach for like 50 years."

The Early Champion

The connectionist heresy had an early champion: Frank Rosenblatt, a psychologist at Cornell. In 1958, he built the Perceptron, a simple machine that could learn to classify patterns by adjusting weighted connections. Show it enough examples of X and not-X, and it would learn to tell the difference.

The New York Times reported breathlessly that the Navy had revealed "the embryo of an electronic computer that it expects will be able to walk, talk, see, write, reproduce itself and be conscious of its existence."

This was hype. But the underlying insight was real. A machine could learn from examples without being explicitly programmed. The weights would adjust automatically. For a brief, intoxicating moment, it seemed like the future of AI.

Then Marvin Minsky and Seymour Papert published *Perceptrons*.

The Proof That Killed

The 1969 book was devastating.

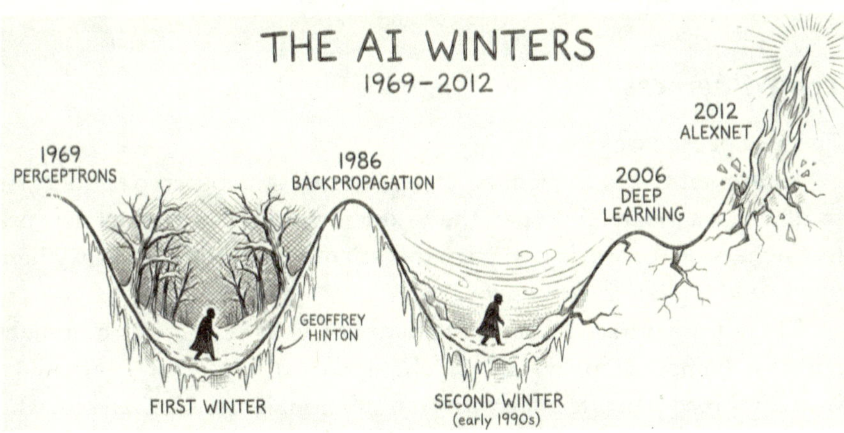

Minsky and Papert were brilliant, rigorous mathematicians. They proved, with formal precision, that single-layer Perceptrons couldn't solve certain fundamental problems. The most famous example: XOR, the exclusive-or function. Even something this simple was beyond the Perceptron's reach.

The proof was correct. The implication, as interpreted by the field, was not.

Minsky and Papert had proven limitations of *single-layer* networks. They speculated, in a few paragraphs, that multi-layer networks might have similar problems, but they didn't prove it. They couldn't. It wasn't true.

But that nuance was lost. The book became an execution. Funding dried up. Students were advised to work on something else. The "AI winter" descended.

"They hadn't actually solved the problem of how to make multi-layer networks learn," Hinton notes, decades later. "They only said they thought this approach wouldn't work."

The thought was enough. For nearly two decades, neural networks were professionally toxic. If you wanted a career in AI, you didn't touch them.

The symbolists had won. Not by proving their approach superior, but by proving the alternative inadequate.

Geoffrey Hinton, in his late twenties, looked at the wreckage and made a choice.

He kept going.

The Wilderness

The next decade was lonely.

Hinton moved to the United States, bouncing between positions. He couldn't get a job in AI proper; the field had no room for neural network researchers. So he worked in cognitive science, in psychology, anywhere that would tolerate his obsession.

There were others. David Rumelhart at UC San Diego, who brought both mathematical rigor and psychological insight. Terry Sejnowski, who combined neuroscience with computation. James McClelland, who studied learning and memory. They formed a loose community, meeting occasionally, sharing ideas, sustaining each other through the exile.

"A whole bunch of us who were working on learning in neural networks," Hinton recalls. "It's something called the PDP group."

PDP: Parallel Distributed Processing. The name captured the core idea – that intelligence emerged from many simple units working simultaneously, with knowledge distributed across connections rather than stored in discrete symbols.

The group's bible would become a two-volume set published in 1986: *Parallel Distributed Processing: Explorations in the Microstructure of Cognition*. It laid out the theoretical foundations for what would eventually be called "deep learning."

But the most important thing Hinton did at San Diego wasn't writing a book.

He solved the problem Minsky and Papert had declared unsolvable.

The Credit Assignment Problem

The core challenge was called the "credit assignment problem."

Imagine a neural network with five layers. Information flows from input to output, passing through layers of neurons. The network makes a prediction. You compare that prediction to reality and discover an error.

Now what?

The error happened at the output. But the mistake might have originated anywhere: in the first layer, the third layer, anywhere the weights are wrong. How do you know which connection to adjust? How do you assign *credit* (or blame) for the error to the right weights?

Single-layer networks, Rosenblatt's Perceptrons, didn't have this problem. There was only one set of weights, so you knew exactly what to adjust.

Multi-layer networks seemed intractable. The error could have come from anywhere. Minsky and Papert's pessimism seemed justified.

The answer turned out to be calculus.

The Algorithm

In 1986, David Rumelhart, Geoffrey Hinton, and Ronald Williams published "Learning representations by back-propagating errors" in *Nature*. It was one of the most important papers in the history of artificial intelligence.

The key insight: use the chain rule.

When the network makes an error, you can calculate how much each weight in the final layer contributed to that error; you just take the derivative. Then, crucially, you can use those derivatives to calculate how much each weight in the *previous* layer contributed. And the layer before that. And so on, all the way back to the input.

The error signal propagates backward through the network, like ripples in reverse. At each step, you adjust the weights slightly in the direction that would have reduced the error. Repeat millions of times. The network learns.

This algorithm, backpropagation, finally made deep networks trainable.

The math wasn't entirely new. Similar ideas had been discovered before, in other contexts. But the 1986 paper crystallized it, demonstrated it on compelling examples, and made it accessible to the field.

"For the first time," historians would later note, "training deep networks seemed possible."

Neural networks were back from the dead.

The Vindication That Wasn't

The 1986 paper generated genuine excitement. For a few glorious years, researchers explored what these newly trainable networks could do. They trained neural networks on pattern recognition, speech processing, simple games. The connectionists emerged from their exile, blinking in the sudden sunlight of funding and attention.

Hinton and Terry Sejnowski developed the Boltzmann Machine, a neural network that borrowed concepts from statistical physics. The learning algorithm was beautiful: the network would "dream," generating patterns from its own internal model, and compare those dreams to reality. The weights would adjust to make the dreams more realistic.

"We were completely convinced that must be how the brain works," Hinton recalls, "and we decided we were going to get the Nobel Prize in Physiology or Medicine."

He laughs when he tells this story now. "It never occurred to us that if it wasn't how the brain works, we could get the Nobel Prize in Physics."

The certainty of youth. The grandiosity that lets you think you'll solve how the brain works in your thirties. Looking back, the Boltzmann Machine was a sidetrack, elegant but impractical. The Nobel would come, eventually, but for different work entirely.

But by the early 1990s, the momentum stalled again.

The networks were painfully slow to train. The computers of the era, Intel 486 processors running at 33 megahertz, couldn't handle large datasets or deep architectures. The networks often got stuck in what mathematicians call "local minima," finding solutions that worked well enough on training data but collapsed when faced with new examples.

Meanwhile, other approaches were working better. Support Vector Machines. Random forests. Ensemble methods. These techniques were mathematically elegant, came with theoretical guarantees about their performance, and consistently outperformed neural networks on the small datasets available.

Machine learning didn't die. But the neural network branch of it was once again out of favor.

A second AI winter descended. Not as harsh as the first. Minsky's *Perceptrons* had been an execution; this was more of a slow freezing. But cold enough. Students were told, again, to work on something else.

"Neural networks 20 or 30 years ago were very primitive in what they could do," Hinton reflects. "They were nowhere near as good as humans at things like vision and language and speech recognition. The idea that you have to now worry about it getting smarter than people, that seems silly then."

This second winter hit differently than the first. In the 1970s, Hinton had been young, building conviction, finding his tribe. By the 1990s, he was middle-aged, the theory proven but the practice stalled. The questions shifted from "Am I right?" to "Will I see it work in my lifetime?"

Most researchers moved on. The pragmatic choice was to work on what worked *now*, not on what might work *eventually* if you believed hard enough.

Geoffrey Hinton moved to Canada.

The Setup

The University of Toronto offered him a professorship in 1987. In Toronto, far from the fashions of Silicon Valley and the politics of MIT, Hinton would build something that looked, from the outside, like a small academic research group.

From the inside, it was something else entirely.

He brought with him the weight of twenty years of fighting for an idea, the scars of two winters, the certainty that the brain held secrets that logic alone couldn't unlock. He brought the memory of Turing and von Neumann, the advocates who died young, and the determination to be the advocate who didn't. He brought the quiet knowledge of what he'd traded for this conviction: the time with his family, the paths not taken, the costs that only showed up in retrospect.

And he brought something else, something that would prove more valuable than any of it.

Students. Young people who believed, as he believed, that the brain held the answer. They would come to Toronto because there weren't many other places to go. They would learn the algorithms, absorb the faith, and carry it forward.

One of them would build AlexNet. Another would build GPT.

But that was twenty-five years in the future.

In 1987, Geoffrey Hinton unpacked his boxes in Toronto and prepared to wait.

It was a monastery. And the monks were about to begin a very long vigil.

6

THE VIGIL
1987–2012

The Refuge

Everyone who mattered in AI was somewhere else. Geoffrey Hinton arrived in Toronto anyway.

Silicon Valley had the venture capital and the startups. MIT and Stanford had the prestige and the government contracts. Japan was pouring billions into its Fifth Generation Computer Project, betting on symbolic AI to dominate the coming century. The future of artificial intelligence, everyone agreed, was being decided elsewhere.

Hinton came to the University of Toronto at age forty, carrying ideas that the mainstream had twice rejected. He was not fleeing in defeat. The backpropagation paper had been published in *Nature* just a year earlier, and there was genuine interest in neural networks. But he sensed what was coming. The excitement would fade. The funding would dry up. The field would move on to something else.

Canada offered something rare: patience.

The Canadian Institute for Advanced Research (CIFAR) had an unusual model. They identified what they called "hard problems," questions that might take decades to answer, that didn't fit neatly into three-year grant cycles. They gave researchers stable support and freedom to follow their intuitions, year after year, without requiring immediate results.

"How does the brain produce intelligence?" qualified as a hard problem.

The funding wasn't lavish. Hinton would never have the computing resources of a Bell Labs or an IBM Research. But he would have something more valuable: time.

He built a lab. He trained students. And he waited.

The Monastery

What Hinton built in Toronto was more than a research group. It was a monastery, a place where a small community of true believers preserved knowledge that the outside world had dismissed.

The students who came through during these wilderness years would later lead the AI revolution. They didn't know that at the time. They just knew they were working on something interesting that nobody else seemed to care about.

The selection pressure was fierce. If you wanted a mainstream career in machine learning, you didn't work on neural networks. You worked on Support Vector Machines, which had mathematical elegance and theoretical guarantees. You worked on ensemble methods, which won competitions. You worked on anything that could demonstrate results *now*.

If you came to Toronto anyway, you were either a true believer or brilliant enough to see something others couldn't.

Often both.

"Because so few people believed in it," Hinton explains, "there weren't many good universities that had groups that did that. So if you did that, the best young students who believed in that came and worked with you. So I was very fortunate in getting a whole lot of really good students."

The luck was real, but it wasn't random. The field's unpopularity filtered for a particular kind of mind: people who could look at the evidence and see something different than the consensus saw, people who could sustain conviction across years of neglect.

The Students

Alex Krizhevsky arrived from Ukraine via Toronto. Quiet, methodical, with an intuitive feel for the computational details that separated working code from broken dreams. He had the patience to optimize systems that others found tedious, the unglamorous work of making things actually run.

Ilya Sutskever came from Israel. Intense, driven, with an almost religious conviction that neural networks were the path to artificial intelligence. Where others saw a research dead end, he saw a sleeping giant. He would become, in Hinton's words, "probably the most important person behind the development of the early versions of ChatGPT."

"I still have lunch with him from time to time," Hinton says. "His parents live in Toronto. When he comes to Toronto, we have lunch together."

The relationships formed in that lab would shape the future of AI. But in the 1990s and early 2000s, that was all invisible. They were just a small group of researchers working on unfashionable problems in a city that wasn't the center of anything.

The outside world had moved on. Inside the monastery, the monks kept copying manuscripts.

The Conspiracy

The Toronto lab wasn't entirely alone. A small circle of believers found each other across institutions and borders.

Yann LeCun, a French researcher, had been working on convolutional neural networks at Bell Labs. These were specialized architectures designed for image processing. His networks could recognize handwritten digits with near-perfect accuracy. AT&T used them to read checks. But the broader AI community dismissed the work as narrow, limited, not the path to general intelligence.

Yoshua Bengio at the University of Montreal was another ally. He and Hinton would trade ideas across the border, two outposts of a scattered resistance.

"For a long time, it was like a small little club," one researcher later recalled. "The three of us would meet at conferences and be the only ones in the room who believed in this stuff."

The group was sometimes jokingly called the "Deep Learning Conspiracy," a handful of researchers keeping the flame alive while the rest of the field moved on to other things. They organized informal workshops. They shared papers that the mainstream ignored. They refined algorithms that nobody was using yet.

In 2018, long after the revolution had vindicated them, Hinton, LeCun, and Bengio would share the Turing Award, the Nobel Prize of computer science. In 2024, Hinton would win the actual Nobel Prize in Physics.

But that was decades away. In the 1990s, they were just the people who hadn't given up on a dead end.

The Waiting

What does it feel like to be right when nobody believes you?

The honest answer is: most of the time, you don't know you're right. You have conviction. You have intuition. You have moments of doubt that you push away. But certainty? That comes later, if it comes at all.

Hinton kept working. New algorithms – the Wake-Sleep Algorithm, Products of Experts, Contrastive Divergence – each a small advance read by a small community of specialists and ignored by everyone else.

"Neural networks 20 or 30 years ago were very primitive in what they could do," Hinton reflects. "They were nowhere near as good as humans at things like vision and language and speech recognition. The idea that you have to now worry about it getting smarter than people, that seems silly then."

The work continued. The winters continued. The waiting continued.

The Missing Piece

The ideas were ready. The algorithms were refined. The theoretical understanding was deep.

What was missing was compute.

Remember the embarrassingly parallel nature of neural networks – the calculations that don't depend on each other? When you train a network, you're doing millions of small operations: multiply this weight by that input, add this bias, apply this function. Each operation is simple. But there are so many of them.

And each operation can be done independently. The multiplication for neuron A doesn't depend on the multiplication for neuron B. You can do them all at the same time.

If you have the hardware.

The CPUs of the era processed instructions sequentially. One calculation, then the next, then the next. Like a single brilliant chess player working through problems one at a time. Training a deep network on a large dataset would take months. Training it on a dataset large enough to demonstrate real capability would take years.

The algorithms worked. They just couldn't work fast enough to prove it.

What was needed was a different kind of processor, one that could do millions of simple calculations simultaneously. Not a single brilliant player, but an army of competent ones.

Such a processor existed. It sat inside every gaming PC.

But Hinton's lab wasn't looking in that direction. They were computer scientists and cognitive psychologists, not video game enthusiasts. The connection between gaming hardware and neural network training wasn't obvious.

Not yet.

The Convergence

In Texas, Carmack was still chasing frame rates. In Santa Clara, Jensen's billion-dollar CUDA bet was finding customers in oil exploration and finance. Neither was thinking about neural networks.

But in early 2007, Nvidia released CUDA to the world. And someone in Hinton's lab made the connection.

The students started experimenting with graphics cards.

"Between about 2006 and 2011," Hinton would later explain at his Nobel lecture, "people were using, particularly in my lab and Yoshua Bengio's lab and Yann's lab, stacks of restricted Boltzmann machines to pre-train feedforward neural networks, and then they would apply backpropagation."

The GPUs made everything faster. Problems that had seemed intractable became manageable. Networks that would have taken months to train could be trained in days.

That same year, Hinton published a paper in *Science* called "Reducing the Dimensionality of Data with Neural Networks." It introduced a technique called "pretraining," a way to initialize deep networks that made them much easier to train. The term "deep learning" started circulating, a rebranding of "neural networks" to escape the stigma of past failures.

Two of his students, George Dahl and Abdulrahman Mohamed, showed that these techniques worked better than anything else for recognizing fragments of speech. The speech recognition community took notice. Hinton's graduates went off to the leading speech groups at Google, Microsoft, IBM.

In 2012, systems based on these techniques went into production at Google. Suddenly, the speech recognition on Android got noticeably better.

Deep learning was working. Not just in the lab, but in the real world.

But it was a competition in October 2012 that would prove decisive.

The Preparation

In 2009, a Stanford researcher named Fei-Fei Li had assembled something unprecedented: a dataset of millions of images, each labeled with what it depicted. Dog. Car. Pizza. Saxophone. Swimming pool. The images came from the internet, labeled by thousands of crowdsourced workers who were paid pennies per classification.

The dataset was called ImageNet. The associated competition, the ImageNet Large Scale Visual Recognition Challenge, invited teams to build systems that could identify what was in an image. Every year, teams would submit their best attempts. Error rates would be compared. Progress would be measured.

For the first few years, traditional computer vision techniques dominated. Hand-crafted features. Support vector machines. Careful engineering of what to look for in an image: edges, corners, textures, shapes. Expert knowledge encoded into expert systems.

The error rates hovered around 26%. Progress was incremental. A percentage point here, half a point there. The problem seemed hard, the improvements slow.

Hinton saw an opportunity.

His students Alex Krizhevsky and Ilya Sutskever had been working on a deep convolutional neural network, a specialized architecture designed for image processing. They had the theory. They had the training algorithm. They had the pretraining techniques.

And now, finally, they had the hardware.

Two Nvidia GTX 580 graphics cards. Consumer gaming equipment. About $500 each.

"We implemented a model with 650,000 neurons and 630 million connections," Krizhevsky would later explain. "We trained it on 1.2 million images from ImageNet."

The training took about a week. The network learned to recognize a thousand different categories of objects, not because anyone had programmed it with rules about what a cat looks like, but because it had extracted patterns from millions of examples.

They called it AlexNet.

They entered the 2012 competition.

The Vigil Ends

In October 2012, Hinton's students submitted their entry to the ImageNet competition.

The demand Carmack had created. The supply Jensen had built. The ideas Hinton had refined for forty years.

Three threads, woven separately, finally crossing.

In a few weeks, the ImageNet results would be announced. The error rate of AlexNet (15.3%) would be so much better than the second-place finisher (26.2%) that it would force the entire field to reconsider everything they thought they knew.

Two $500 graphics cards. A week of training time. The "impossible" done on $1,000 of consumer gaming hardware.

The vigil was ending.

And the world was about to change.

7

TWO GRAPHICS CARDS

The Dinosaur Bones

Data was dinosaur bones. That's how Fei-Fei Li remembers it. While other computer scientists refined their algorithms, polishing mathematical techniques like jewelers working gemstones, Li stood in conference halls arguing that they had the whole thing backward.

"Data doesn't matter," the establishment said. What mattered was elegant math. Clever features. Human-designed rules for telling a cat from a dog.

Li thought they were blind. She had a PhD in electrical engineering from Caltech, a background in computational neuroscience, and an insight that seemed obvious to her and crazy to everyone else: the reason computers couldn't see wasn't that the algorithms were wrong. It was that the algorithms were starving.

Human children don't learn to recognize faces by studying twenty photographs. They don't master language by memorizing grammar rules. The human visual system is trained on millions of images, a relentless, continuous stream of visual data during the early years of development. The brain extracts patterns from abundance. It learns what matters by exposure to everything.

Why should artificial vision be any different?

"The limiting factor," Li argued, "isn't the sophistication of the algorithms. It's the poverty of the data."

The computer vision community was not persuaded. Data, they told her, was simply the substrate on which algorithms ran. A utility. A resource to be catalogued, like museum specimens.

Like dinosaur bones.

"While data is now considered the 'new oil,'" Li would later reflect, "in 2009 it was viewed as dinosaur bones – curiosities to be catalogued but not useful fuel for the engine of technology."

Li decided to prove them wrong. She would build the largest, most diverse, most carefully labeled collection of images ever assembled. She would give the algorithms what they needed: not elegant math, but an ocean of data so vast that patterns would emerge from sheer abundance.

The calculation was staggering. WordNet, the linguistic database she would use as her taxonomy, contained roughly 80,000 noun categories. If she wanted 500 to 1,000 clean, labeled images for each category, she needed tens of millions of images.

She tried hiring undergraduate students. The math was brutal: at the rate undergrads could work, labeling images for $10 an hour, the project would take ninety years.

Ninety years.

Then a graduate student mentioned something called Amazon Mechanical Turk.

"He showed me the website," Li recalled, "and I can tell you literally that day I knew the ImageNet project was going to happen."

The 48,000

Mechanical Turk was a marketplace for micro-tasks: tiny jobs that humans could do in seconds but computers couldn't do at all. Label this image. Transcribe these words. Click yes or no.

Li's team reverse-engineered the problem of building a visual cortex for computers. They scraped 160 million candidate images from the internet. They designed tasks so simple that anyone could do them: here's an image, here's a definition ("German Shepherd: a breed of large shepherd dogs used in police work"), click yes or no.

They paid fractions of a cent per task. They required multiple workers to agree before accepting any label. They built consensus mechanisms, quality filters, statistical checks.

And then they unleashed it on the world.

Forty-eight thousand workers from 167 countries participated. At peak efficiency, the collective workforce labeled tens of thousands of images per day. The average worker could evaluate fifty images per minute. The costs ran into tens of thousands of dollars, a fortune for an academic lab, pocket change for the scale of the achievement.

By 2009, Li's team had assembled 3.2 million labeled images across 5,247 categories. By 2010, the full ImageNet contained over 14 million images across 21,841 categories.

It was, by orders of magnitude, the largest labeled image dataset ever created.

And almost nobody cared.

The Poster Session

The 2009 Conference on Computer Vision and Pattern Recognition was held in Miami. Li and her team submitted their paper, "ImageNet: A Large-Scale Hierarchical Image Database," expecting, perhaps, a keynote slot. A plenary session. Recognition of what they'd built.

They got a poster session.

Not even an oral presentation. A poster, in a humid exhibition hall, competing for attention with dozens of other posters while conference attendees wandered past on their way to more prestigious events.

Li stood there, sweating in the Miami heat, explaining her dataset to anyone who stopped. Her team handed out ImageNet-branded pens, trying to attract attention.

The reception was skeptical at best. "If you can't even do one object well," one researcher asked, "why would you do thousands?"

This was the prevailing wisdom. The path to computer vision ran through better algorithms, not bigger data. You solve the small problem first, get one object recognition working really well, and then scale up.

Li's approach seemed backward. Ambitious to the point of absurdity. A massive collection of labeled images that no existing algorithm could exploit.

The poster session ended. The pens ran out. The computer vision community went back to refining their elegant mathematics.

But Li had built something. And in three years, someone would use it to change the world.

The Competition

Li did what visionaries do when ignored: she created a forcing function.

Starting in 2010, the ImageNet Large Scale Visual Recognition Challenge gave researchers something to compete for. A curated subset of the data: 1.2 million images across 1,000 categories. Dogs (dozens of breeds), cats (many varieties), cars, planes, mushrooms, coffee cups. The task was simple to state: given an image, identify what's in it.

The metric was "top-5 error rate." Your system got five guesses, and if the correct answer appeared anywhere in those guesses, you got credit. The distinction between a Siberian Husky and an Alaskan Malamute is real but semantic; ImageNet was hard enough without penalizing systems for reasonable confusion.

The competition attracted the best teams in computer vision. They brought the best tools the field had developed: Fisher Vectors, SIFT descriptors, Support Vector Machines. Human experts designed features by hand, then trained classifiers to use those features.

Year after year, the error rates stayed stuck around 26%.

The dinosaur bones sat there, catalogued and labeled, waiting for someone to realize they weren't bones at all.

They were fuel.

The Hardware Constraint

In Toronto, Krizhevsky faced a problem.

The network he was building had 60 million parameters. It needed to store those parameters, plus the activations from each layer, plus the gradients for backpropagation, plus the input images themselves. Each GTX 580 had 3 gigabytes of memory.

Three gigabytes wasn't enough.

A single card couldn't hold the network.

So Krizhevsky did something that would later become standard practice: he split the network across both GPUs. Certain layers lived on one card, certain layers on the other. The two GPUs communicated at specific synchronization points, shared just enough information to maintain coherence, then returned to processing their portions independently.

This wasn't in any textbook. Neural networks were supposed to be unified systems, not bisected artifacts running on separate chips. But Krizhevsky had a problem to solve and hardware that couldn't solve it the normal way.

The split worked. The network trained. What had seemed like a limitation became the first instance of "model parallelism," a technique that would become essential when models grew to billions of parameters.

In 2012, it was just a hack to get around memory limits.

But hacks, sometimes, change the world.

The Dataset Meets the Network

The dinosaur bones, it turned out, were exactly the fuel these algorithms needed.

The Toronto team took an approach that would have seemed absurd to the 2009 conference skeptics. They showed the network raw pixels. No hand-designed features. No human expertise about what to look for. Just 1.2 million images and their labels, fed through 60 million trainable parameters.

The network would figure out what mattered.

The architecture drew on techniques refined over decades. Dropout, randomly disabling neurons during training, to prevent overfitting. ReLU activation functions – "rectified linear units," a deceptively simple rule that outputs zero for negative inputs and passes positive inputs unchanged – to avoid the vanishing gradient problem. Data augmentation, translating, reflecting, and color-shifting images, to make 1.2 million examples feel like more.

None of these ideas were new. What was new was the scale. What was new was the hardware that made scale possible.

The training took about a week. Images in, errors measured, gradients propagated backward, weights adjusted. Repeat. Millions of times.

On September 30, 2012, the team submitted their predictions to the ImageNet evaluation server.

And then they waited.

The Nobel Prize That Wasn't

Remember the Boltzmann Machine story? Hinton and Sejnowski, convinced they'd cracked how the brain works, certain a Nobel Prize in Physiology or Medicine awaited them?

"It never occurred to us," Hinton laughs now, "that if it wasn't how the brain works, we could get the Nobel Prize in Physics."

In 2024, he did get the Nobel Prize in Physics, for foundational contributions to machine learning using artificial neural networks. The physics prize, not medicine. The irony is perfect.

But in 2012, waiting for ImageNet results, no prize was in sight. Just decades of work, an unfashionable field, two students with a network running on gaming hardware, and a competition that might vindicate everything or prove nothing.

The Waiting

What does it feel like to be right when nobody believes you?

Hinton had some experience with this. Forty years of believing something the field had twice rejected. The long decades while the mainstream moved on.

You maintain conviction through doubt. You train students who will carry the work forward. You refine algorithms that nobody uses, publish papers that barely get cited, attend conferences where your talks draw sparse audiences.

And sometimes, after forty years, you turn out to be right.

The ImageNet results would be announced in October, at the European Conference on Computer Vision in Florence. Researchers from around the world would gather to see whose methods had won, whose techniques had advanced the field, whose careful engineering had squeezed out another fraction of a percent.

The Toronto team had entered something different. Not a refinement. Not an increment.

A demolition.

8

THE ANNIHILATION

Florence

October 2012. A conference room in Italy.

The European Conference on Computer Vision had become a fixture of the academic calendar, a gathering where researchers presented their latest work, argued about methods, sized up the competition. The ImageNet challenge was one highlight among many: a benchmark, a measuring stick, a way to track how much progress the field had made.

The room wasn't expecting a revolution.

The slide appeared. Three columns: team name, method, error rate.

ILSVRC 2012

ImageNet Large Scale Visual Recognition Challenge

Team	Method	Top-5 Error
SuperVision	Deep CNN	15.3%
ISI	Fisher Vectors	26.2%
OXFORD_VGG	Fisher Vectors + SVM	26.9%

10.8% ahead!

Someone in the audience inhaled sharply.

That couldn't be right. The gap was too large. Previous years had seen improvements of one or two percentage points, the gradual accumulation of refinements, the slow grind of scientific progress. This was something else.

10.9 percentage points.

The second-place finisher used the best of traditional methods: Fisher Vectors, Support Vector Machines, hand-crafted features refined over decades of research. They achieved 26.2%, a respectable result, consistent with the field's trajectory.

The Toronto team had beaten them by the combined progress of five years. In a single entry.

Some researchers assumed there must be an error. The evaluation server had malfunctioned. The submission was invalid. Something.

There was no error.

Two graphics cards. About a week of training. A network that learned its own features instead of being told what to look for.

The impossible had been done on $1,000 of consumer gaming hardware.

The Reaction

Yann LeCun was in the audience.

He had spent thirty years on convolutional neural networks, the same architecture, fundamentally, that AlexNet used. He had trained networks to read handwritten digits on checks. He had published papers that were cited by dozens, not thousands. He had been part of the small conspiracy of believers who kept faith through the winters.

He called it a "turning point."

The younger researchers understood immediately. One result doesn't overturn a field, but this wasn't one result. This was proof of concept at scale. Everything the neural network believers had claimed was now demonstrated, undeniably, on the hardest benchmark the computer vision community had.

IMAGENET 2012 RESULTS:
AlexNet: 15.3%
Runner-Up: 26.2%
Third: 27.1%

The methods that seemed promising in September were obsolete in October.

"It was like suddenly everyone realized they'd been doing it wrong," one researcher later recalled. "All that careful engineering, all those clever tricks – none of it mattered. You just throw pixels at a network and it figures out what's important."

The older researchers were more cautious. Neural networks remained "black boxes." The mathematical guarantees that made Support Vector Machines theoretically elegant were absent. Where was the explanation? Where was the understanding?

But the numbers didn't lie. 15.3%. Nothing else came close.

The paradigm had shifted.

What AlexNet Learned

The network's internal representations became almost as famous as the result.

Krizhevsky analyzed what the different layers had learned. The first layer, the one that directly processed raw pixels, had discovered something that looked like neuroscience.

Edge detectors at various orientations. Color blob detectors. The same features that neuroscientists had found in the primary visual cortex of cats and monkeys.

Nobody had told the network to look for edges. The training signal – "this is a dog, this is a cat, this is a mushroom" – contained no information about edges. The network discovered that detecting edges was useful for the classification task, and so it evolved edge detectors through backpropagation.

The subsequent layers built on these primitives. Layer two combined edges into textures and simple patterns. Layer three detected parts of objects: eyes, wheels, handles. The later layers combined parts into wholes.

"And about 12% of ImageNet is dogs," Hinton noted. "And to be good at ImageNet, you have to tell the difference between very similar kinds of dog."

The network had become an expert in fine-grained classification. Not because anyone programmed dog-recognition rules, but because the training process shaped sixty million parameters into a system that could distinguish a Golden Retriever from a Yellow Labrador from a lion.

Feature learning, not feature engineering. The network discovered what mattered. The humans just provided the examples.

The Cathedral

There's an old story about blind men and an elephant. Each man touches a different part – trunk, leg, ear, tail – and describes a different animal. The trunk feels like a snake. The leg feels like a tree. None of them perceive the whole.

Only from a distance, looking back, does the elephant become visible.

The gaming industry spent three decades creating a market for massively parallel processors. This was done to make entertainment better. Teenagers buying graphics cards to play *Counter-Strike* were voting with their dollars for hardware development.

The chips that emerged, thousands of cores capable of trillions of floating-point operations per second, turned out to be almost perfectly suited for training neural networks. This was not the design goal. It was an accident of mathematics.

The same operations that render a frame of *Quake* also train a layer of AlexNet: matrix multiplications, millions of them, done in parallel. The kid complaining about frame rates in 1996 was creating market pressure for exactly the hardware that would train ChatGPT in 2022.

The neural network researchers spent decades developing algorithms that required exactly this kind of hardware. This was not coordination. Hinton didn't know Jensen. Jensen didn't know Carmack. Carmack had never thought about backpropagation.

But the consumer market, driven by entertainment, funded R&D that would have been impossible to justify for AI research alone. The market for fun became the market for intelligence.

The Connection Complete

Return, for a moment, to the specific.

The GTX 580 graphics card that Krizhevsky used was designed for gaming. Its specifications were chosen to optimize frame rates in first-person shooters. Its marketing materials talked about anti-aliasing and texture filtering, not machine learning.

But inside that card was an architecture shaped by thirty years of demand for parallel processing. Thousands of simple cores. High memory bandwidth. The ability to do millions of independent calculations simultaneously.

This architecture emerged because programmers like Carmack kept pushing the boundaries of what real-time graphics could do. Each new game demanded more. Each new demand shaped the next generation of hardware.

Jensen recognized that this architecture had uses beyond gaming. CUDA let anyone program the GPU directly – not just for graphics, but for any parallel calculation. This was a bet, and like all bets, it might have failed.

It didn't fail. Krizhevsky wrote CUDA kernels. The kernels trained AlexNet. AlexNet won ImageNet.

The thread runs unbroken.

Carmack wanted to make demons run faster → gamers demanded better graphics → Nvidia built parallel processors → Jensen bet on CUDA → Krizhevsky wrote GPU code → Hinton's algorithms finally had hardware → AlexNet proved deep learning worked.

Each link in the chain required the previous links. Remove any one, and the chain breaks.

This is the deepest lesson of the convergence: no one planned it.

The transformation that will reshape human civilization emerged from a video game company in Texas, a chip manufacturer in California, and a professor in Toronto, none of whom knew they were working on the same project.

The Convergence

Three threads.

The demand Carmack created. The supply Jensen built. The algorithms Hinton refined.

In October 2012, the threads crossed. The gaming hardware, funded by the demand for better graphics, proved Hinton's forty-year bet was right.

And yet looking back, it seems almost inevitable. The market for entertainment funded infrastructure that happened to be perfect for intelligence. The algorithms that needed hardware found hardware that needed algorithms. The pieces fit because the math was the same: parallel operations, matrix multiplications, the deep structure of computation that doesn't care whether you're rendering demons or recognizing dogs.

Two graphics cards. Three researchers. One competition.

The annihilation of everything that came before.

The beginning of everything that came after.

INTERLUDE

THE COFFEE SHOP THAT LEARNED

The story pauses here.

You've just witnessed the annihilation – AlexNet's victory in October 2012. Two graphics cards, sixty million parameters, a gap so large the room gasped. But what actually happened? How did a neural network defeat a decade of conventional wisdom?

This interlude explains the technology behind the triumph. No equations required – just a coffee shop, some workers, and the question of blame.

When we return, the builders say their farewells.

Backpropagation: The trick that made deep learning possible

You've seen how backpropagation solved the credit assignment problem – how Hinton and his colleagues figured out which weights to adjust, by how much, in which direction. The math involved derivatives and chain rules.

But what does it actually *feel* like? Here's an intuition.

The Coffee Shop

Imagine a coffee shop. Not a Starbucks with its automated precision, but something more chaotic, a place where your drink passes through four people before it reaches your hand.

Sam picks the beans. Jordan grinds them. Taylor controls the water temperature. Alex does the pour.

Each worker is like a neuron in a neural network: a small processing unit that takes an input, does something to it, and passes the result along.

You take a sip. It's bitter. Too bitter.

Now here's the credit assignment problem: *Who do you blame? And how much?*

The Blame Problem

Your instinct might be to blame Alex, since he's the last one who touched the drink. But Alex was working with whatever Taylor handed him. And Taylor was dealing with Jordan's grind. And Jordan was stuck with Sam's bean choice.

Everyone shares some responsibility. But how much? Should Alex adjust 50%? Should Sam change 2%? And in which direction: should Jordan grind finer or coarser?

Each worker has settings they can adjust: how fine to grind, what temperature to use, how long to pour. In neural network terminology, these adjustable settings are called *weights* or *parameters*. A network might have thousands or millions of these little knobs that could be turned up or down.

Getting the final answer wrong was easy to detect. Figuring out which knobs to adjust, by how much, and in which direction – that was the puzzle that had stumped the field for decades.

The naive approach: freeze everyone else, wiggle one person's technique, make a new coffee, taste it. Now you know that person's effect. Repeat for everyone.

Four coffees just to calculate one round of adjustments. Now imagine your chain has a thousand workers. That's a thousand coffees to figure out one set of corrections. Then you adjust, and need another thousand for the next round. Impossibly slow.

The Trick

What Hinton and his colleagues showed in 1986 was that you didn't need to make all those coffees. You could calculate everyone's blame in a single backward sweep, no extra testing required.

Making a coffee, passing beans through the chain until you have a drink, is what AI researchers call a *forward pass*. You only need one forward pass to make the coffee. The magic of backpropagation is that you only need one backward pass to calculate all the blame.

The insight relied on derivatives. If you remember anything from calculus, a derivative just measures how fast something is changing. When you're driving and glance at your speedometer, you're looking at a derivative: how fast your position is changing at that exact moment.

Applied to our coffee shop, you ask derivative questions: If Alex poured one second longer, how much more bitter would it taste? If Taylor made the water one degree hotter, how would that change what Alex had to work with, which then changed the taste?

This blame signal, how much each worker contributed to the error and in which direction, is called the *gradient*. It tells you both the magnitude (how much to adjust) and the direction (which way to turn the knob).

Here's the key. Sam's blame isn't measured directly. It's calculated by multiplying through the chain:

Sam's effect on Jordan × Jordan's effect on Taylor × Taylor's effect on Alex × Alex's effect on taste

Each step multiplies. If each link in the chain is a small number, say 0.3, then Sam's total blame might be 0.3 × 0.3 × 0.3 × 0.3. About one percent. That's why Alex might need to adjust his weights by 50% while Sam adjusts only 2%. Not because anyone guessed. Because the math demanded it.

Which Way to Turn the Knob

But magnitude is only half the answer. You also need direction. Should Jordan grind finer or coarser?

This is where the sign of the derivative comes in. When you ask "if Jordan grinds finer, does the coffee get more bitter or less bitter?" the answer isn't just a number. It's positive or negative.

Positive means: grinding finer makes it worse. So grind coarser.

Negative means: grinding finer makes it better. So grind finer.

The gradient gives you both pieces, magnitude and direction, in one calculation. No guessing. Just measurement, multiplication, and adjustment.

The Dance

There's one more subtlety. You don't let anyone fix 100% of their blame in one go.

Why? Because when Alex changes, it shifts what the right answer is for Taylor. And Taylor's change affects what Jordan should do. If everyone overcorrects at once, the next coffee might be worse than the first.

So everyone nudges just a little. This nudge size is called the *learning rate*: too big and the system overcorrects; too small and learning takes forever. Finding the right learning rate is part of the art.

Make a coffee. Taste it. Calculate the gradient. Nudge all the weights. Make another coffee. Taste. Calculate. Nudge.

Thousands of times. Sometimes millions. Each cycle is called a *training step*.

And here's something that surprises people: Jordan might grind coarser in round one, then finer in round fifty. The right answer for Jordan depends on what everyone else is doing. As they move, Jordan's target moves. It's like five people balancing on a wobbly platform together. You shift left, I shift right, then you shift back, I shift back too.

Eventually, if things go well, the system settles. Everyone finds a configuration where the coffee tastes good and nobody's blame is significant anymore. The platform stops wobbling. That's *convergence*. That's a trained neural network.

The Resurrection

The 1986 paper was a direct answer to Minsky and Papert. Multi-layer networks *could* be trained. The blame *could* be assigned. The math *wasn't* intractable – you just had to multiply derivatives backward through the chain.

XOR? Trivial. The same networks that Minsky and Papert had declared hopeless could now learn it in seconds.

The winter began to thaw. But a new problem emerged, one that would take another twenty years to solve.

The Deep Problem

Backpropagation worked beautifully for short chains. Four workers? No problem. Eight? Still fine.

But the real power of neural networks was supposed to come from *depth*: many layers, each building more abstract representations than the last. The first layer might detect edges. The second, shapes. The third, objects. The fourth, scenes. The more layers, the more sophisticated the understanding.

The trouble was, deep chains broke the blame game.

Imagine your coffee shop grows. Now there are twenty stations between the beans and your cup. Twenty workers, each making small adjustments to what they receive before passing it along.

You taste the coffee. Bitter.

You start calculating blame backward. Worker twenty gets a clear signal. Worker nineteen, slightly weaker. Worker eighteen, weaker still. By the time the signal reaches worker one, the person picking the beans, it has been multiplied through nineteen intermediate steps.

Remember: 0.3 × 0.3 × 0.3 × 0.3 was already just one percent. Now multiply twenty times.

The gradient doesn't just shrink. It *vanishes.*

Worker one receives blame so microscopic it might as well be zero. They never learn. They keep picking the same beans, round after round, while only the last few workers in the chain ever improve.

This was called the *vanishing gradient problem*. And it meant that deep networks, the kind that could truly think in layers of abstraction, were essentially untrainable. Backpropagation had answered Minsky and Papert, but it couldn't deliver on the deepest promise of neural networks.

Hinton spent years on this problem. So did others. The field made progress on shallow networks, but the deep architectures that seemed most promising remained out of reach.

The Apprenticeship

In 2006, Hinton published a paper that changed the game: "Reducing the Dimensionality of Data with Neural Networks."

The insight was deceptively simple. If the blame signal couldn't survive the journey from end to beginning, don't send it all at once. Train each station *separately first*.

Back to our twenty-station coffee shop. Instead of connecting everyone on day one and hoping the blame signal survives, you start differently.

Worker one, the bean selector, trains alone. They're not trying to make good coffee yet. They're just learning to recognize patterns in beans: which ones are similar, which are different, what features matter. They develop expertise in their own domain, without knowing anything about the final product.

Then worker two trains. They receive worker one's output and learn to recognize patterns in *that*, the ground coffee. They develop their own expertise.

And so on, station by station, each worker becoming competent in their domain before being connected to the chain.

Hinton called this *pre-training*. Each layer learns useful representations on its own, unsupervised, without needing to know the final answer.

Only after everyone has developed basic competence do you connect the full chain and *fine-tune* with backpropagation. Now the gradient doesn't need to teach from scratch. It just nudges workers who already know their craft. Small adjustments to their weights, not wholesale learning.

The vanishing gradient doesn't vanish as badly when you're fine-tuning experts instead of training novices.

For the first time, networks with many layers could actually learn. Deep learning, the term Hinton championed, was finally possible.

The Inspection Line

There was one more insight, and it came not from Hinton but from a young French researcher at Bell Labs.

Yann LeCun had a different problem. He wasn't trying to recognize abstract patterns; he was trying to read handwritten zip codes. The U.S. Postal Service processed millions of letters daily. Humans had to squint at each envelope, deciphering the scrawl. LeCun thought neural networks could help.

But images are strange inputs. A photograph might contain a million pixels. If you connected every neuron to every pixel, the standard approach, the network would have billions of parameters. Far too many to train. Far too many to even store in memory.

LeCun's insight, published in his 1989 paper "Backpropagation Applied to Handwritten Zip Code Recognition": *don't look at the whole image at once.*

Return to our coffee shop. Word spreads. The shop gets successful. Now they source beans from a hundred different farms, samples arriving every morning in a long line. The owner needs to evaluate them all quickly.

You can't build a hundred separate Sam-Jordan-Taylor-Alex chains. Too expensive. Too many workers. Too many parameters to manage.

So the owner reorganizes.

Maria is the Acidity Inspector. She walks to sample #1, takes a sip, writes "3.2–moderately acidic." Steps to sample #2. Same technique. Same evaluation. She walks the entire line, leaving a trail of acidity scores behind her.

Tomás follows as the Body Inspector. Same process: one skill, applied everywhere.

Then *Rosa* for sweetness. *Diego* for aftertaste.

Each inspector applies *one skill* across all hundred samples. Same technique everywhere, *same weights*. Maria doesn't retrain her palate at each sample. She uses the same acidity-detection method from start to finish.

The trail of scores Maria leaves behind, that's called a *feature map*. A record of what one skill detected across the whole line.

Then a supervisor combines their reports. She looks at sample #47: high acidity from Maria, heavy body from Tomás, low sweetness from Rosa, clean finish from Diego. "That's a natural-process Brazil," she says. "Buy it."

Four local scores, combined into recognition.

LeCun called this a *convolutional neural network*. The name comes from the mathematical operation, sliding a small filter across an image, but the intuition is simple: scan locally with shared weights, combine globally.

He used it to read zip codes. Then bank checks. The networks were small by later standards, thousands of parameters, not millions, but they worked. A machine could look at handwriting and see numbers.

The same vanishing gradient problem plagued LeCun's networks. Deep convolutions meant many layers. Many layers meant disappearing blame signals. He got them working, but only barely, only on small images, only with patience and tricks.

What LeCun needed was what Hinton needed: more compute.

The Final Piece

By 2006, the theoretical pieces were in place. Backpropagation told you how to calculate gradients. Pre-training let you build deep networks. Convolutions let you process images efficiently with shared weights.

But there was still a bottleneck. Training a deep convolutional network on a large dataset, millions of images, say, took weeks or months on the computers of 2006. Every experiment was a major investment. Iteration was painful. Ideas that should take hours to test took forever.

Hinton had the algorithm. LeCun had the architecture. What they lacked was the computational power to prove it could scale.

They would wait three more years for gaming hardware to catch up.

The Third Heretic

There was a third researcher in this stubborn cohort. Yoshua Bengio, working in Montreal, was focused on a different question: not images, but language.

How do you train a neural network to predict the next word in a sentence? To understand that "the cat sat on the ___" probably ends with "mat" or "chair" but not "democracy"?

In 2003, Bengio published "A Neural Probabilistic Language Model," a paper that would become foundational for everything that followed in natural language processing. Where LeCun taught machines to see, Bengio was teaching them to read.

The same computational bottleneck applied. Language models needed to process vast amounts of text, adjusting millions of parameters through countless training steps. It was agonizingly slow.

Hinton, LeCun, Bengio. Three researchers, three problems – structure, vision, language – all waiting for the same thing.

We will return to Bengio's contribution later, when the story shifts from images to words. For now, know that when the hardware finally arrived, it wouldn't just unlock vision. It would unlock everything.

~

In 2009, in Toronto, Alex Krizhevsky would plug a GeForce GTX 280 into his gaming computer and discover that training which took three days on a CPU took four hours on a GPU.

The algorithm that answered Minsky and Papert. The convolutional architecture that learned to see. The pre-training trick that unlocked deep networks. The language models waiting in the wings. And now, the hardware that could run them at scale.

Four breakthroughs, spanning forty years, finally converging.

The second AI winter was about to end.

A Translation Guide

For readers who encounter AI terminology elsewhere, here's how our coffee shop maps to the technical language:

Neuron A worker in the coffee shop. A small processing unit that takes an input, transforms it, and passes the result along.

Layer A station in the chain, or in the inspection line, a group of specialists all doing the same type of evaluation.

Weights / Parameters Each worker's technique settings. How fine Jordan grinds. What temperature Taylor uses. How long Alex pours. Maria's calibrated palate for acidity. The adjustable knobs.

Activation What gets passed to the next station. The intermediate product: ground beans, heated water, the almost-finished drink.

Forward Pass Making a coffee. Running inputs through the chain from beginning to end.

Loss / Error How bad the coffee tastes. The gap between what you got and what you wanted.

Gradient The blame signal. How much each worker contributed to the error, and in which direction they should adjust.

Backpropagation Calculating blame backward through the chain. Multiplying derivatives from end to beginning.

Learning Rate How much each worker nudges their technique per round. Too big and you overcorrect; too small and learning takes forever.

Training Step One cycle of: make coffee, taste, calculate blame, nudge all the weights.

Convergence When the coffee tastes good and no one needs to adjust anymore. The system has settled.

Vanishing Gradient When the blame signal shrinks to nothing in long chains. Workers at the beginning never learn because they can't hear the feedback.

Pre-training Training each worker separately before connecting the chain. Building expertise station by station.

Fine-tuning Connecting pre-trained workers and making small adjustments with backpropagation. Nudging experts rather than teaching novices.

Convolutional Neural Network (CNN) The inspection line approach. Specialists who walk the line with one skill, leaving a trail of scores, then combining reports into recognition.

Feature Map The trail of scores Maria leaves behind. What one inspector detected across the whole line.

Shared Weights Maria uses the same acidity-detection technique at every sample. One set of parameters applied everywhere, reducing what the network needs to learn.

Now you know how they learn. The coffee shop is no longer a mystery – and neither is the technology reshaping your world.

9

THE BUILDERS, AFTERWARD

The Auction

The phone started ringing almost immediately.

Google. Microsoft. Baidu. Facebook. Every major technology company with any interest in machine learning wanted the team that had built AlexNet. Not just the code – the code would eventually be published – but the people. The expertise. The intuition for what worked.

Deep learning talent was suddenly the scarcest resource in technology.

Hinton had a problem, though. A personal problem that shaped what happened next.

"I have a son who has learning difficulties," he explained years later. "And in order to be sure he would never be out on the street, I needed to get several million dollars."

Academia couldn't provide that. He had tried teaching a Coursera course, hoping the online education boom might generate serious money. "But there was no money in that."

The only path to financial security was selling himself to a company. And fortunately, he had just demonstrated something that companies desperately wanted.

"So when I was 65, fortunately for me, I had two brilliant students who produced something called AlexNet, which was a neural net that was very good at recognizing objects and images."

Hinton, Krizhevsky, and Sutskever incorporated a company – DNNresearch Inc. – that had no products, no business plan, no assets except the employment contracts of its three founders.

Then they held an auction.

"We actually set up an auction where we had a number of big companies bidding for us."

This wasn't a normal acquisition. There was nothing to acquire except the right to hire three researchers. But the bidding was intense. Baidu sent representatives. Microsoft competed. Google brought Larry Page.

Google won. The reported price was in the tens of millions of dollars – some sources say $44 million – for a shell company whose only value was the brains of three people.

The acquisition signaled something larger than any individual career move. Deep learning wasn't a research curiosity anymore. It was strategic. Major corporations would pay tens of millions to acquire capability.

The talent wars had begun.

The Conversion

In 2012, AlexNet was the only neural network entry in the ImageNet competition.

In 2013, almost every entry was a neural network.

The conversion was nearly instantaneous. Researchers who had spent careers on traditional computer vision pivoted within months. Graduate students abandoned thesis projects. Entire research groups changed direction.

The phase transition was complete. Before AlexNet: hand-crafted features, elegant mathematics, slow progress. After AlexNet: learned features, brute-force optimization, exponential improvement.

The techniques would prove general. If neural networks could recognize objects in images, perhaps they could recognize words in speech. Perhaps they could translate between languages. Perhaps they could play games at superhuman levels.

But those discoveries belong to the next act.

The Three Depart

What became of the builders?

Carmack had already left id Software in 2013, joining Oculus to work on virtual reality. He spent nearly a decade pushing the boundaries of what headsets could do – latency, rendering, the hard problems of immersion. Then, in 2022, he announced he was leaving to focus full-time on artificial general intelligence.

The man who made demons run faster now wanted to make them think.

He had seen, finally, where all his work had led. The parallel processing philosophy he had pioneered for games had become the foundation for machine intelligence. The demand he had created had funded the hardware. The hardware had enabled the algorithms. And now the algorithms were doing something that looked, increasingly, like thought.

Carmack wanted to be part of what came next.

Jensen rode a wave he never saw coming.

Nvidia's market capitalization grew from roughly $10 billion in 2012 to over $4 trillion by December 2025, making it the most valuable company on Earth. The gaming company had become the infrastructure provider for the AI revolution.

He still wears the leather jacket. He still wakes up afraid. The survival instinct that kept Nvidia alive through the graphics card wars now serves him in a different kind of competition – the race to supply the computational foundation for artificial intelligence.

The bet on CUDA, which seemed like an expensive distraction in 2006, turned out to be the bet that made everything else possible. Jensen had built infrastructure for a future he couldn't see. The future had arrived.

Hinton won – and then he worried.

In early 2013, he joined Google Brain. He would work there for the next decade, watching neural networks go from vindicated research to dominant paradigm. The Turing Award came in 2018, recognition from the field that had rejected his ideas for so long. The Nobel Prize in

Physics came in 2024, recognition that the ideas had become foundational science.

But somewhere along the way, triumph curdled into concern.

Hinton left Google in 2023. Not because he was angry with the company. Because he wanted to speak freely about danger.

He had realized something that shook him: these systems would eventually surpass human intelligence. And humanity had never been second before.

The godfather of AI became its most prominent critic. The man who kept faith through forty years of rejection now warns that his victory delivered more than he bargained for.

"They're going to get smarter than us," he says, "and we don't know how to control them."

This, too, is part of the story. The cathedral builders didn't just construct a building. They built something that might outlive the need for builders.

The Handoff

The three who built the foundation now step back from the center of this story.

Carmack had created the demand. Jensen had built the supply. Hinton had refined the algorithms. Their threads had crossed in October 2012, and that crossing was complete.

But the cathedral they constructed would be occupied by others. The next decade belongs to the scalers – researchers with conviction, companies with resources, believers who would push the architecture further than its creators imagined.

Ilya Sutskever, the quiet student who co-authored AlexNet, would co-found OpenAI and become its chief scientist. Alex Krizhevsky, the methodical Ukrainian who wrote the CUDA kernels, would work at Google before stepping back from the spotlight entirely. Yann LeCun, the French researcher who had kept faith alongside Hinton, would become Chief AI Scientist at Meta – a role he would hold for over a decade before striking out on his own.

New names would emerge: Sam Altman, Demis Hassabis, Dario Amodei. New companies would form: DeepMind, Anthropic, xAI. New architectures would arrive: the transformer, the large language model, the systems that could not just see but speak.

The builders will return. They have opinions about what's being built in their cathedral – strong opinions, divergent opinions, warnings and hopes and fears. We will hear from them again in the final act.

But for now, the story follows the explosion.

The Renaissance begins.

ACT II: THE EXPLOSION

THE DECADE OF SCALE (2012–2022)

The unplanned infrastructure finds its purpose. Deep learning goes from curiosity to dominant paradigm. The Renaissance gains momentum – and the tools that will eventually reach everyone begin to take shape.

A Note on the Players

Act II introduces many of the key figures who built the AI revolution. To help readers track the rapidly shifting landscape, names are tagged with their primary affiliation – the company or model most associated with their work.

Format: ***Name*** *(Company) or* ***Name*** *(Company→Company) for those whose movement is part of the story. For example: Ilya Sutskever (OpenAI→Safe Superintelligence) shows a key figure whose departure became part of the narrative.*

Note: Facebook rebranded as Meta in 2021; the current name is used throughout.

In an industry where talent moves fast and allegiances shift faster, these tags capture where each builder made their mark.

10

THE SCALERS

The Conviction

Ilya Sutskever (OpenAI→Safe Superintelligence) knew something the field hadn't accepted yet.

The algorithms existed. The datasets were growing. The compute was becoming available. Geoffrey Hinton had been publishing papers about neural networks since the 1980s. The math hadn't changed.

What was missing was something harder to quantify.

"The ideas were all there," Sutskever would later explain. "The thing that was missing was a lot of supervised data and a lot of compute. Once you have a lot of supervised data and a lot of compute, then there is a third thing which is needed as well, and that is conviction."

Conviction that if you took the right stuff, which already existed, and applied it with enough data and enough compute, it would work.

This sounds obvious now. It wasn't obvious then.

Ilya Sutskever had the conviction.

He'd had it since 2010 or 2011, when he connected two facts in his mind. The first was that neural networks could be trained with ten layers. James Martens had demonstrated this with the Hessian free optimizer, training a deep network end-to-end without the tricks that earlier researchers had needed. The second was simpler: ten layers was enough.

"Because if you have a neural network with 10 layers, it's as though you allow the human brain to run for some number of milliseconds. Neuron firings are slow, and so in maybe 100 milliseconds, your neurons only fire 10 times. So it's also kind of like 10 layers."

In 100 milliseconds, you can perfectly recognize any object.

If the brain could do it with ten layers of neural firing, then a neural network with ten layers should be able to do it too. You just needed to train it properly. You needed the data. You needed the compute.

"And when that happened," Sutskever recalls of the Hessian free optimizer moment, "I thought, this is it."

The Student and the Teacher

Sutskever had arrived in Hinton's lab at the University of Toronto as a master's student in 2003, then stayed for his PhD. He was young, barely twenty, and had grown up in Jerusalem before his family moved to Canada. His parents were engineers. He was drawn to mathematics, to physics, to the question of how minds worked.

Hinton's group in Toronto was small. The monastery had few monks. But the ones who came were true believers.

"People who worked in machine learning simply didn't think that neural networks could do much," Sutskever explains. "People didn't believe that large neural networks could be trained."

The establishment had moved on. Support Vector Machines were elegant. Ensemble methods worked reliably. Neural networks were a dead end from the 1980s; everyone knew this. The smart money was elsewhere.

Sutskever wasn't interested in the smart money.

He was interested in the question that had obsessed Hinton for forty years: How does meat think? And he was increasingly convinced that the answer was scaling. Make the networks bigger. Give them more data. Let them learn.

The 2012 ImageNet result proved they were right. Two graphics cards. A week of training. The gap between AlexNet and the competition wasn't a few percentage points – it was ten. The paradigm shifted overnight.

Sutskever was one of the three authors. He was twenty-seven years old.

The Talent Wars

The AlexNet auction set off a war.

Deep learning talent was suddenly the scarcest resource in technology. The 2013 ImageNet results showed that AlexNet wasn't a fluke – neural networks dominated again. The 2014 results showed it wasn't even close. By 2015, ResNet achieved error rates below human performance on the benchmark.

The companies responded the only way they knew how: by acquiring capability.

Facebook hired Yann LeCun – one of the original heretics, the inventor of convolutional networks who had been vindicated alongside Hinton – to run a new research division called FAIR. Google, not content with the Toronto acquisition, bought DeepMind for over $500 million in January 2014. The London-based company had barely demonstrated anything publicly, but its team of researchers had convinced Google that they were building something transformative.

The talent war drove salaries into the stratosphere. Fresh PhD graduates in deep learning were commanding offers of $1 million or more. Companies bid against each other for researchers who could publish influential papers, who could train larger models, who had the conviction.

Yoshua Bengio (Mila, the Montreal AI research institute he founded), the third member of what Yann LeCun called "the Deep Learning Conspiracy," stayed in Montreal. He built an ecosystem instead of joining a corporation. His lab would produce a generation of researchers, many of whom would eventually disperse to the companies competing for talent.

The concentration had begun.

Only well-funded organizations could afford to participate. Training runs cost tens of millions of dollars. The scaling hypothesis, the emerging belief that bigger was simply better, meant that progress would require resources that only corporations could provide.

This wasn't how science usually worked. Academic labs had driven the field for decades. Now the center of gravity was shifting to Mountain View and Seattle and London.

The heretics had won. But their victory came with consequences.

What Ilya Believed

Sutskever's conviction was more specific than the general enthusiasm for deep learning.

He believed in scale.

Not just that neural networks worked; everyone accepted that by 2013. But that they would continue to work, better and better, if you just made them bigger. That there was no ceiling. That the same architecture that recognized dogs in photographs might, with enough parameters and enough data, do almost anything.

"The key fact about deep learning before deep learning started to be successful is that it was underestimated," Sutskever explains. "People didn't believe that large neural networks could be trained."

Now that belief was wrong. The question was: how large? How far did the scaling go?

The pattern was emerging empirically. Larger models performed better – not a little better, but measurably, consistently better across tasks. The improvements were predictable. You could draw curves. You could extrapolate.

This was strange. Most engineering involves diminishing returns. Make a bridge twice as heavy and you don't get a bridge twice as strong. Make a car engine twice as big and you don't get twice the power. Complexity introduces inefficiencies.

Neural networks seemed different. Double the parameters, double the data, and you got predictable improvements. The curves didn't bend.

If this was true – if scale really was all you needed – then the implications were staggering. You didn't need new algorithms. You didn't need fundamental breakthroughs. You just needed more of everything.

More data. More compute. More patience.

And more conviction that the curves would hold.

March 2016: Move 37

The general public noticed something was happening in March 2016.

The game was Go, an ancient board game, deceptively simple in its rules but astronomically complex in its possibilities. A 19×19 grid. Black and white stones. Players take turns placing stones, trying to surround territory. Capture your opponent's stones by completely encircling them.

Chess has roughly 35 legal moves from any given position. Go has roughly 250. A chess game lasts about 40 moves on average. A Go game lasts about 200. The total number of possible Go games exceeds the number of atoms in the observable universe.

For decades, AI researchers had believed Go would resist computers far longer than chess. When Deep Blue beat Kasparov in 1997, experts predicted it would be decades before a computer could beat a professional Go player. The game required intuition that couldn't be captured in explicit rules.

DeepMind's AlphaGo proved them wrong.

In March 2016, AlphaGo played a five-game match against Lee Sedol, one of the greatest Go players in history, a nine-dan professional with eighteen world championship titles. The match was broadcast live. Millions watched.

AlphaGo won the first two games.

Then, in game two, something strange happened.

Move 37. AlphaGo played a stone in a position that no human would consider. The commentators were baffled. The experts couldn't understand it. The move violated basic Go intuition, the kind of knowledge that professional players absorbed over decades of study.

Lee Sedol left the room for fifteen minutes.

"It was a fantastic, but also somewhat bittersweet moment," Demis Hassabis, DeepMind's co-founder, would recall. He understood what had happened. AlphaGo had played millions of games against itself. It had explored possibilities that humans had never considered. It had found something new.

Move 37 was extrapolation. Not just pattern matching, not just interpolation from human examples, but a creative discovery – something no human had taught it, something it had found on its own by playing the game at a scale no human could match.

The machine had done something beautiful. And the world noticed.

AlphaGo won the match four games to one. The single game Lee Sedol won, game four, came after he played his own surprising move, move 78, that briefly confused the system. It was a reminder that humans still had tricks. But the overall message was clear.

Deep learning wasn't just for image recognition anymore.

The Scaling Hypothesis

The AlphaGo victory crystallized something that the researchers already suspected.

Neural networks weren't narrow tools. They weren't pattern matchers limited to one domain. Given enough data, enough compute, enough training, they could master anything humans could formalize.

Go was supposed to be different from chess. Go required intuition. Go required creativity. Go required judgment that couldn't be reduced to brute-force calculation.

AlphaGo hadn't used brute force. Not exactly. It had used deep neural networks to evaluate positions – to develop intuition through experience, the same way humans did. It just had more experience than any human could accumulate in a lifetime.

The scaling hypothesis wasn't formally stated until years later. But its outlines were visible by 2016:

Larger models perform better. The improvement is predictable. There is no obvious ceiling.

If this was true, then the path forward was clear. You didn't need to solve the alignment problem. You didn't need to understand how the networks represented knowledge. You didn't need fundamental breakthroughs in architecture or training.

You needed resources.

And OpenAI was founded to acquire them.

The Speciesist

The origin of OpenAI was a fight about what it meant to be human.

Elon Musk (xAI) and Larry Page (Google) had been friends for years. Musk had known Page since before Google got venture funding, back when it was just a search engine that might work. They stayed at each other's houses. They talked late into the night about the future.

The future they talked about most was artificial intelligence.

"I was arguably the prime mover behind OpenAI," Musk would later explain, "in the sense that it was created because of discussions that I had with Larry Page back when he and I were friends and stayed at his house and talked to him about AI safety."

The conversations were not going well, according to Musk. As he described it, Page's position was philosophical: if AI surpassed human intelligence, why should that be a problem? The disagreement ran deep – Musk would later recount being called a "speciesist" for prioritizing human welfare over digital intelligence.

Meanwhile, Google had acquired DeepMind. They had probably two-thirds of all the AI researchers in the world. They had essentially infinite money and compute.

Musk decided something had to be done. If Google was going to pursue AGI without concern for human welfare, there needed to be a counterweight. A nonprofit, open-source, transparent alternative.

He started recruiting.

The Recruiting Battle

The linchpin was Ilya Sutskever.

Everyone knew it. If you wanted to build AGI, you needed the best researchers. And Sutskever – Hinton's student, AlexNet co-author, one of the true believers who had proven that neural networks worked – was among the best.

He was at Google Brain. Google didn't want to lose him.

"That was a tough recruiting battle," Musk recalled. "It was mostly Demis on one side and me on the other, both trying to recruit Ilya."

Demis Hassabis, co-founder of DeepMind, now part of Google. One of the most respected AI researchers in the world. He wanted Sutskever (OpenAI→SSI) to stay in the Google ecosystem.

Musk wanted him for the new nonprofit.

"Ilya went back and forth. He was going to stay at Google, then he was going to leave, then he was going to stay, then he was going to leave."

The decision wasn't just about salary or resources. Sutskever had access to Google's compute, Google's data, Google's talent pool. The new organization had nothing but a mission and some funding.

But the mission mattered.

"Finally, he did agree to join OpenAI. That was one of the toughest recruiting battles we've ever had."

The choice said something about Sutskever's values. He believed that AI development needed to happen in the open, with safety as a priority. He believed that concentration of power, even in a company he respected, was dangerous. He believed that someone needed to build AGI the right way.

"That was really the linchpin for OpenAI being successful," Musk said.

The friendship between Musk and Page did not survive.

"Really the breaking of the friendship was over OpenAI. And specifically, I think the key moment was recruiting Ilya."

The Nonprofit with Ambition

OpenAI announced itself in December 2015 with a billion-dollar commitment from Elon Musk (xAI), Sam Altman (OpenAI), and other investors.

Musk provided over $40 million of the initial funding. The name was his idea too – the "open" in OpenAI was supposed to mean open source, a transparent alternative to the closed development happening at Google and elsewhere.

The stated mission was breathtaking: ensure that artificial general intelligence benefits all of humanity. Not narrow AI. Not chatbots or image classifiers. AGI – the kind of intelligence that could match or exceed human capabilities across any domain.

Most AI researchers thought this was decades away at minimum. Some thought it was impossible. The announcement read like science fiction.

But the team was serious.

Sutskever became chief scientist. He had traded the security of Google's resources, the stability of working for one of the most profitable companies in history, for a nonprofit startup with no products, no revenue, and a mission that sounded like hubris.

"You know, it's crazy, that all of this is real," Sutskever would later reflect. "Like, all this AI stuff and all this Bay Area, yeah, that it's happened, like... Isn't it straight out of science fiction?"

He believed. That was why.

He believed that the scaling hypothesis would hold. He believed that the techniques that had won ImageNet and mastered Go would continue improving. He believed that OpenAI, with enough resources and the right team, could push the frontier.

The bet was on scale. More parameters. More data. More compute. More conviction.

And the bet was on doing it differently than Google would.

The Philosophical Commitment

The bet on scale was more than a resource allocation decision. It was a philosophical commitment.

The AI establishment, what remained of it, still believed that progress required fundamental breakthroughs. New algorithms. New architectures. New insights into how intelligence worked. The history of AI was littered with promising approaches that had hit ceilings: expert systems that couldn't scale, reasoning engines that couldn't handle uncertainty, symbolic AI that couldn't learn from data.

The scalers said something different. They said: the algorithm already exists. We just need more of everything.

This was heresy of a different kind than Hinton's original neural network faith. Hinton had believed in neural networks when everyone else believed in symbols. The scalers believed that the existing neural networks, the ones that already worked, would become intelligent if you just made them bigger. No new ideas required. Just compute.

"So in terms of deep learning, to answer the question directly," Sutskever explained, "the ideas were all there."

All there. Since the 1980s. Since Rumelhart and Hinton and Williams published their paper on backpropagation. The math hadn't changed.

What changed was conviction. And resources.

The Concentration

By 2016, the consequences of the scaling hypothesis were becoming visible.

Training a state-of-the-art neural network required millions of dollars in compute. The datasets were proprietary – scraped from the internet by companies with the engineering resources to do so at scale. The talent was expensive; the best researchers commanded salaries that only corporations could afford.

Academic labs, which had driven AI research for decades, found themselves increasingly marginalized. They couldn't afford the training runs. They couldn't attract the talent. They published papers, but the cutting-edge work happened behind corporate doors.

The scaling hypothesis centralized AI research. It shifted power from universities to corporations. It created a new class of researcher, the empiricist with access to a cluster, and devalued the theorist working with pencil and paper.

This worried some people. OpenAI's nonprofit structure was supposed to be a hedge – a way to pursue AGI outside the profit motive, for the benefit of humanity rather than shareholders.

But nonprofits need resources too. And the scaling hypothesis demanded resources that only certain actors could provide.

The cathedral was being built. But only the well-funded could afford to work on it.

What Hinton Thought

Hinton, meanwhile, was settling into Google Brain.

He worked three days a week. His back problems made sitting difficult, so he avoided flying and stood most of the time. He supervised research, advised on projects, thought about the problems that had occupied him for forty years.

He wasn't worried yet.

The systems were impressive, but they were narrow. AlexNet could recognize dogs. AlphaGo could play Go. They couldn't do both. They couldn't do the million other things that a human child learns in the first few years of life.

The gap between what the networks could do and what humans could do remained vast. True intelligence – the kind that could adapt to any situation, reason about anything, understand the world the way humans understood it – seemed distant.

The vindication was real. The Nobel Prize would come eventually. But the danger? The existential risk? That still felt like science fiction.

"Neural networks 20 or 30 years ago were very primitive in what they could do," Hinton would later reflect. "They were nowhere near as good as humans at things like vision and language and speech recognition. The idea that you have to now worry about it getting smarter than people, that seems silly then."

It seemed silly.

The scaling curves hadn't bent yet.

2017

By 2017, the foundation was almost complete.

AlexNet had proven that deep learning worked. AlphaGo had proven it could master domains that required intuition. The scaling hypothesis had emerged from empirical observation into something like natural law. The talent wars had consolidated the best researchers into a handful of organizations. OpenAI had assembled a team and a mission.

One piece was missing.

The architecture that dominated deep learning – the convolutional neural network, the recurrent neural network – had been developed decades earlier. They worked well enough. But they had limitations.

Recurrent networks struggled with long sequences. They couldn't pay attention to distant parts of the input. They processed information sequentially, one step at a time.

The scaling hypothesis said bigger was better. But some architectures scaled better than others.

In June 2017, a team at Google published a paper with an almost arrogant title: "Attention Is All You Need."

They had invented something new.

They called it the transformer.

11

THE FOUNDATIONS

The Architecture

Eight Google researchers proposed throwing away everything that worked. Keep only the attention mechanism. Discard the rest.

Previous architectures used attention as a supplement, an addition to the recurrent or convolutional structures that formed the backbone of deep learning. This paper, published in June 2017, said: throw away the backbone.

They called it the "transformer."

The architecture processed sequences in parallel rather than sequentially. It could examine an entire sentence at once, beginning to end, without working through the words step by step.

The key innovation was "self-attention." Each element of the input could attend to every other element, weighted by learned relevance. When processing the word "it" in a sentence, the network could learn to focus on whatever "it" referred to, no matter how distant.

Previous architectures struggled with long sequences. Recurrent neural networks degraded over distance; by the end of a long paragraph, information from the beginning had weakened, the signal fading as it passed through sequential steps.

Transformers didn't have this problem. Every position could attend to every other position directly. Distance didn't matter.

The paper demonstrated state-of-the-art results on machine translation. The improvements were substantial but not revolutionary. It read like solid incremental progress.

The consequences would take years to unfold. By the time they did, most of the authors would have left Google.

The Inventors

They came from different teams, brought together by a problem: machine translation was important to Google's business, and existing approaches had limitations.

The key insight came from Ashish Vaswani and Noam Shazeer – that attention mechanisms, used in isolation, might be powerful enough to replace the recurrent architectures everyone assumed were necessary.

Shazeer had joined Google in 2000, employee number twenty-something, when the company was small enough that everyone knew each other's names. He had worked on search, on ads, on the infrastructure that made Google run. He had watched the company grow from a startup in a garage to one of the most valuable corporations on Earth.

He had also been thinking about AI since before Google existed.

"A friend of mine at the time had told me that his New Year's resolution for 2000 was to live to see the year 3000," Shazeer recalls, "and that he was going to achieve this by inventing AI. So I was like, oh, that sounds like a good idea."

Twenty-five years later, he was still working on it.

The transformer paper represented the distillation of years of work on attention mechanisms, sequence modeling, and the properties of deep learning at scale. The authors knew they had something good. They didn't know they had something transformative.

"We did need to see the size of the opportunity," Shazeer would later admit, "often reflected in what other companies are doing."

The Exodus

The paper was published openly. Google didn't keep it secret. This was standard practice for academic research: publish your findings, advance the field, build your reputation through contribution.

It was also, in retrospect, one of the most consequential giveaways in the history of technology.

Within two years, OpenAI had used the transformer architecture to build GPT. Within five years, ChatGPT had reached 100 million users. Competitors created tens of billions of dollars in market value using the architecture that Google researchers invented.

The transformer authors didn't stay at Google.

Noam Shazeer (Google→Character) left to found Character.AI, a company that let users create and chat with AI personas. "The main reason I left Google to start Character was I thought that the biggest thing that the LLM industry could use was an application where anybody can go and interact with LLMs," he explained. "This was before ChatGPT launched, before Gemini launched."

He saw the opportunity. Google didn't ship it.

Others scattered to different startups and competitors. The team that invented the foundation of modern AI dispersed across the industry, taking their expertise with them.

Google had become the Xerox PARC of AI: inventing transformative technology and watching others commercialize it.

The Organizational Failure

Why didn't Google capitalize on its own invention?

The company had everything: the talent, the hardware, the data, the architecture. Google's TPUs were custom-built for training neural networks. Google's data centers held the scale of compute that the scaling hypothesis demanded. Google's research teams had produced breakthrough after breakthrough.

But Google was also a search company.

"Our view of things from a search perspective," Jeff Dean (Google) would later acknowledge, "was like these models hallucinate a lot and they don't get things right correctly a lot of the time. And that means that they aren't as useful as they could be."

From Google's perspective, a chatbot that sometimes fabricated information was dangerous. Search was supposed to provide correct answers. Users trusted Google to be authoritative. A language model that confidently stated false information could damage that trust.

"We'd like to make that better. And from a search perspective, you want to get the right answer 100 percent of the time, ideally going to be very high on factuality. And these models were not near that bar."

The safety concerns were real. The hallucination problem was real. Google's caution wasn't unreasonable.

But it meant that Google had an internal chatbot system, operational before ChatGPT launched, that employees could use during the pandemic. Google didn't ship it.

"I think what we kind of didn't quite appreciate," Dean admitted, "was how useful they could be for things you wouldn't ask a search engine. Like help me write a note to my veterinarian or can you take this text and give a quick summary of it."

The technology was ready. The organization wasn't.

The GPT Progression

While Google hesitated, OpenAI executed.

The transformer architecture gave them the foundation. The scaling hypothesis gave them the strategy. Ilya Sutskever (OpenAI→SSI), as chief scientist, oversaw the progression.

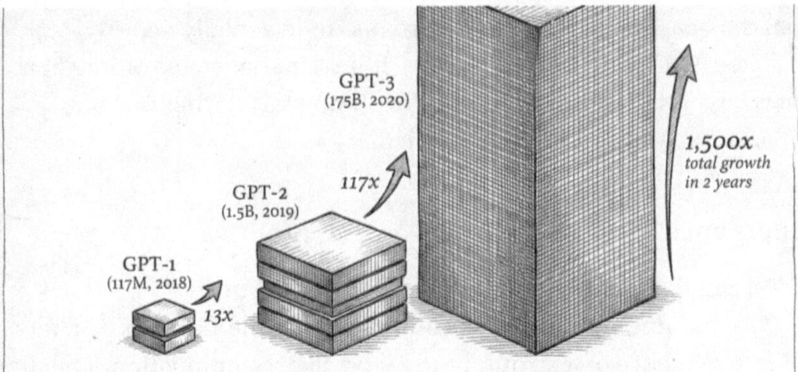

THE SCALING FAITH: GPT PARAMETER GROWTH

GPT-1 arrived in June 2018 – a transformer with 117 million parameters, trained on books. It demonstrated that pre-training on large amounts of text could transfer to downstream tasks. Train a language model on raw text, then fine-tune it for specific applications.

This was the "generative pre-training" that gave GPT its name. The model learned patterns from reading, then applied those patterns to new situations.

By February 2019, **GPT-2** had grown to 1.5 billion parameters – more than ten times larger. The results were startling.

GPT-2 could generate coherent paragraphs of text. Given a prompt, it would continue writing in a plausible style. It could complete stories, generate fake news articles, produce dialogue that sounded human. The quality was inconsistent (sometimes brilliant, often nonsensical) but the flashes of capability were undeniable.

OpenAI made a controversial decision: they announced the model but didn't release the full version.

"Too dangerous to release," they said.

The claim generated skepticism. Was this genuine concern about misuse, or was it a marketing stunt? The AI research community debated. Some accused OpenAI of overhyping a text generator. Others took the safety concerns seriously.

Either way, GPT-2 demonstrated something important: language models were becoming good enough that their outputs could be mistaken for human writing. The Turing test was becoming less relevant than the question of what these systems could actually do.

Then **GPT-3**, in May 2020: 175 billion parameters. More than 100 times larger than its predecessor. The jump was enormous.

And something strange happened.

Emergence

GPT-3 could do things it wasn't explicitly trained to do.

The training objective was simple: predict the next word. That's all. Given a sequence of text, output the most likely continuation. The model had no explicit instruction in arithmetic, translation, programming, or reasoning.

But GPT-3 could perform all of these tasks, imperfectly and inconsistently, but recognizably.

Give it a few examples of addition problems, and it would solve new ones. Give it a few examples of English-to-French translation, and it would translate new sentences. Give it a few examples of Python code, and it would generate new functions.

This was "few-shot learning." Instead of training on thousands of examples, you showed the model a handful of demonstrations in the prompt, and it figured out the pattern.

Nobody had programmed this capability. It emerged from scale.

The researchers at OpenAI noticed it first in small experiments. Someone would try the model on a task it had never been trained for, and it would work. Not perfectly, but recognizably. The Slack channels filled with screenshots. *Did you see what it did with this prompt?* The capability had emerged from the training, unplanned and unexpected.

"So when you don't know anything," Sutskever explained the phenomenon, "you'll notice very broad strokes, surface level patterns... Eventually, you may start to notice that there are certain words that occur often. You may notice that spellings are a thing. You may notice syntax. And when you get really good at all these, you start to notice the semantics. You start to notice the facts."

The larger the model, the more it noticed. GPT-2 noticed syntax. GPT-3 noticed facts. GPT-3 noticed that addition problems had patterns. GPT-3 noticed that translations followed structures.

The scaling curves were holding. Bigger models performed better. And the improvements weren't just quantitative; they were qualitative. New capabilities emerged at scale that simply didn't exist in smaller models.

"The key fact about deep learning before deep learning started to be successful," Sutskever reflected, "is that it was underestimated."

It was still being underestimated.

The Turing Award

In March 2019, the Association for Computing Machinery announced its Turing Award recipients.

Geoffrey Hinton. Yann LeCun (Meta). Yoshua Bengio (Mila).

Hinton was in his Toronto office when he got the call. After four decades of pushing an idea that most of the field had dismissed, the highest honor in computer science was his to share. LeCun heard the news in New York. Bengio in Montreal.

The three godfathers of deep learning. The heretics who had kept faith through the AI winters. The conspiracy that had become the establishment.

The Turing Award is computer science's highest honor, often called the Nobel Prize of computing. It came with a $1 million prize, shared among the three.

The citation praised their "conceptual and engineering breakthroughs that have made deep neural networks a critical component of computing."

For Hinton, it was vindication after forty years. The ideas he had developed as a young researcher – backpropagation, Boltzmann machines, the belief that neural networks could learn – had finally been recognized as foundational science.

For LeCun, it was recognition of his persistence. Convolutional neural networks, developed in the 1980s, had been ignored for decades before AlexNet proved they could scale.

For Bengio, it was validation of his choice to remain in academia – building an ecosystem in Montreal rather than joining a corporation, training the students who would disperse across the industry.

The heretics had won. The conspiracy had prevailed.

But there was something bittersweet about the timing. The award came in 2019, when the revolution was still accelerating. The transformer had been published. GPT-2 had been announced. GPT-3 was in training.

The three laureates were being honored for work that had started decades ago. Meanwhile, the systems their ideas had spawned were growing more powerful by the month.

Hinton wasn't worried yet. Not really. The systems were impressive, but they were narrow. They couldn't match human capability across the full range of cognition. The gap remained large.

The worry would come later.

The Jensen Thread

Throughout this decade, from AlexNet in 2012 to GPT-3 in 2020, one constant held.

Every breakthrough ran on Nvidia hardware.

The transformer trained on GPUs. GPT-2 trained on GPUs. GPT-3, with its 175 billion parameters, required clusters of V100s, thousands of them working in parallel.

DeepMind's AlphaGo? GPUs. DeepMind's AlphaFold, which solved the protein folding problem in 2020? GPUs. Every major AI lab, every scaling run, every training cluster. GPUs.

Jensen Huang's bet from 2006 was paying off beyond imagination.

When he invested in CUDA, he couldn't have known that deep learning would explode. When he told his board that the gaming market was important but there might be something bigger, he didn't know that something bigger would be artificial intelligence.

But the architecture matched the need. The parallel processing that rendered video games also trained neural networks. CUDA meant that researchers could write efficient training code without designing custom hardware. The gaming market funded R&D that the AI market couldn't have justified.

Nvidia's market capitalization grew from roughly $10 billion in 2012 to over $300 billion by the end of 2020. The growth was accelerating. The company that made chips for gamers was becoming the infrastructure provider for artificial intelligence.

They had all been building the same cathedral. And Jensen, without planning it, had supplied the tools.

2020

By the end of 2020, the foundation was complete.

The transformer architecture existed, invented at Google, adopted everywhere. The scaling hypothesis had been validated: larger models performed better, reliably, predictably. The talent had consolidated into OpenAI, into DeepMind, into the corporations that could afford the training runs. The hardware existed. Nvidia's GPUs, optimized over a decade for parallel computation.

GPT-3 could write essays, generate code, answer questions, translate languages. It could do things it wasn't trained to do. It exhibited capabilities that emerged from scale, capabilities that surprised even the people who built it.

The researchers knew what was coming. The public didn't.

Two years until ChatGPT.

The Question Nobody Could Answer

But there was a question that nagged at the researchers.

What did the model actually understand?

GPT-3 could generate grammatically correct sentences about quantum mechanics. Did it understand quantum mechanics? It could write code that compiled and ran. Did it understand programming? It could produce text that sounded like it came from a human. Was it intelligent?

The critics said no. "Just" statistical pattern matching, they argued. The model predicted likely words based on training data. It didn't understand anything; it was a sophisticated autocomplete.

The defenders said the distinction was getting fuzzy. If the model could answer questions correctly, solve problems, generate novel text that humans found useful, what more did you need to call it understanding? At what point did sufficiently sophisticated pattern matching become intelligence?

The honest answer was that nobody fully knew.

"I don't want to imply that our models have complete semantic understanding because that's not true," Sutskever acknowledged. "But they definitely are showing signs of semantic understanding, partial semantic understanding, but the smaller models do not show those signs."

Partial. Signs. The language was careful.

The scaling laws suggested that if you made the models bigger, they would get better still. The curves hadn't bent. The improvements continued. GPT-4 was already in development.

But understanding? Consciousness? Intelligence in the way humans meant it?

Those questions remained open.

What Was Missing

The technology was ready.

GPT-3 existed. It could generate text that impressed and occasionally astonished. It could help with coding, writing, brainstorming. It could answer questions and complete tasks. The capability was real.

But the capability was locked away.

Access to GPT-3 required an API key. Developers could build applications on top of it, but ordinary users couldn't just talk to the model. The interface was technical, developer-focused, not designed for mass adoption.

The scaling hypothesis had produced powerful systems. The organizational structures had produced powerful capabilities. What was missing wasn't technology.

What was missing was a product.

An interface. A way to put this capability in the hands of ordinary people.

That would come in November 2022.

And when it did, a hundred million people would meet their first AI.

12

THE INTERFACE

The Demo That Escaped

November 30, 2022. A Wednesday. Nothing special about the date.

OpenAI pushed a new product to the web. They called it ChatGPT. It was, technically, nothing new: GPT-3.5 under the hood, the same model that had been available through their API for months. What was new was how you accessed it. A simple chat window. Type a question. Get an answer.

No API key. No developer documentation. No technical knowledge required.

The product team at OpenAI had been testing various interfaces for months. The chat format emerged almost by accident, a demo that proved unexpectedly compelling. Sam Altman's philosophy guided the decision: "We need to deploy to learn." Rather than wait for perfection, they shipped.

Within five days: one million users.

Within two months: one hundred million users.

Nothing in the history of consumer technology had grown this fast. Not Facebook, which took four and a half years to reach that milestone.

Not Instagram, which took two and a half years. Not TikTok, which took nine months.

ChatGPT did it in sixty days.

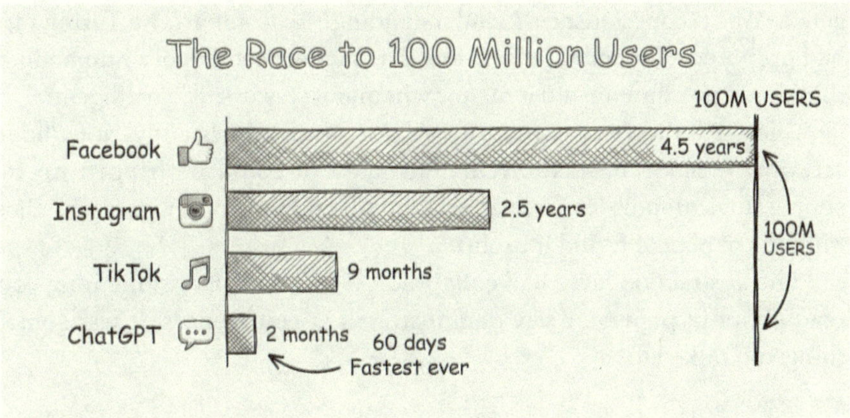

"The striking thing about the reaction to ChatGPT is not just the number of people who are blown away by it, but who they are," Paul Graham of Y Combinator tweeted that December. "These are not people who get excited by every shiny new thing. Something big is happening."

The Abstraction Collapsed

For years, AI had lived behind walls.

GPT-3, released in 2020, required an API key. Developers could build applications on top of it, but ordinary users couldn't just talk to the model. DALL-E generated images, but access was gated. The capability existed, but it was locked away, accessible only to those who knew how to request it programmatically.

ChatGPT demolished the wall.

The interface was almost embarrassingly simple. A text box. A submit button. A response that streamed in word by word. It looked like every messaging app people already used: iMessage, WhatsApp, Slack. There was no learning curve. You didn't need to understand transformers or attention mechanisms or token limits. You just typed.

And the system responded.

"If you can't tell the difference," Sam Altman (OpenAI) observed in interviews, "how much do you care?"

The question cut to the heart of what was happening. For decades, philosophers and computer scientists had debated the nature of intelligence. Was it consciousness? Understanding? Reasoning? The Turing test had proposed a functional definition: if a machine could fool a human into thinking it was human, did it matter whether it was "really" intelligent?

ChatGPT didn't pass the Turing test – not consistently, not under scrutiny. It made mistakes. It hallucinated. It could be tripped up by simple logical puzzles. But it was good enough, often enough, that millions of people found it useful.

The abstraction layer had collapsed. AI was no longer something you read about in papers or saw demonstrated at conferences. It was something you talked to.

The Screenshots as Marketing

The virality was self-generating.

People discovered things ChatGPT could do and shared screenshots on social media. A lawyer asked it to draft a motion. A teacher asked it to explain quantum physics to a fifth grader. A programmer asked it to debug code. A novelist asked it to continue a story.

The screenshots spread because the conversations were surprising. The AI didn't just answer questions; it engaged. It adopted styles. It apologized when wrong. It asked clarifying questions. The responses felt, for lack of a better word, human.

"I got ChatGPT to..." became a meme format. I got ChatGPT to write a breakup text in the style of Shakespeare. I got ChatGPT to explain the Federal Reserve like I'm five. I got ChatGPT to write code for a Chrome extension.

Each screenshot was an advertisement that cost OpenAI nothing. Users were marketing the product to each other, driven by genuine astonishment.

The conversation shifted overnight. Before ChatGPT, AI discussions happened in academic conferences and technical forums. After ChatGPT, AI discussions happened in teachers' lounges and law offices and

Thanksgiving dinners. The vocabulary entered common speech. "ChatGPT" became a verb.

The Operator

Sam Altman wasn't a researcher. He wasn't an engineer, not in the way that Ilya Sutskever (OpenAI→Safe Superintelligence) or John Carmack were engineers. He was something else: an operator.

He had run Y Combinator, the startup accelerator that had funded companies like Dropbox, Airbnb, and Stripe. He understood how to build organizations, how to raise capital, how to position products. When OpenAI was founded in 2015, Elon Musk and others brought the vision. Altman brought operational expertise.

"I think about AGI, about artificial general intelligence," Altman said in interview after interview. "That's what we're building."

The ambition was visible in every conversation. Other AI labs talked about narrow applications: better search, improved image recognition. OpenAI talked about intelligence itself. The mission statement was breathtaking: "Ensure that artificial general intelligence benefits all of humanity."

Most AI researchers thought AGI was decades away at minimum. Some thought it was impossible. The claim read like science fiction.

But the team was serious. And ChatGPT was proof that they could deliver.

"My kids will never be smarter than AI," Altman reflected. "They will never grow up in a world where products and services aren't as smart as they are."

This wasn't a prediction about the distant future. It was a statement about the present, about what had already begun.

The Deployment Philosophy

Altman's approach to AI development differed from the traditional academic model. In academia, you published papers. You advanced the state of the art on benchmarks. You competed for citations.

OpenAI had published papers too, the GPT papers, the DALL-E

papers. But Altman pushed for something more: products. Deployments. Systems that ordinary people could use.

"We need to deploy to learn," was the philosophy. You couldn't understand how people would use AI by theorizing about it in a lab. You had to ship.

The approach generated controversy. Critics argued that OpenAI was moving too fast, releasing capabilities without fully understanding the risks. The hallucination problem was real: ChatGPT could generate confident assertions that were simply false. The jailbreaking community found ways to bypass safety guardrails. The potential for misuse was obvious.

But Altman argued that controlled deployment was better than no deployment. If AI was going to be powerful, the public needed time to adjust. Institutions needed time to develop policies. The alternative, keeping capability locked away until some hypothetical moment of safety, meant the transition would be sudden rather than gradual.

"People are smart, society is resilient," Altman maintained. "I think people are quickly understanding that this is now a thing."

Whether this philosophy was wisdom or recklessness would only become clear in retrospect. What was undeniable was that it worked. ChatGPT reached more users faster than any AI product in history.

The Education Panic

Within weeks of launch, schools were in crisis.

Students discovered that ChatGPT could write essays. Not rough drafts – polished essays, with thesis statements and supporting arguments and proper citations. The writing wasn't always correct, and it wasn't always good, but it was often good enough to pass.

Teachers panicked. How do you assign homework when students can have an AI do it? How do you grade essays when you can't be sure who wrote them? How do you teach writing when the fundamental incentive structure has collapsed?

Some schools banned ChatGPT entirely. New York City public schools blocked access on their networks. Others tried to implement AI detection tools, which proved unreliable. Some educators argued for a different approach: embrace the technology, teach students to use it critically, adjust the curriculum.

The debate echoed earlier panics, about calculators, about Google, about Wikipedia. Each technology had prompted fears about student learning, about what skills would become obsolete, about what it meant to be educated. Each time, the educational system had adapted.

But this felt different. Calculators performed arithmetic. Wikipedia provided information. ChatGPT could generate text that sounded like thinking.

The Hallucination Problem

The technology was not without flaws.

ChatGPT hallucinated. It would make claims with complete confidence that were simply false. Ask it about a historical figure and it might invent biographical details. Ask it for legal citations and it might generate case numbers that didn't exist. Ask it about current events and it might describe things that hadn't happened.

The hallucination problem was inherent to the architecture. Language models predicted the most likely next word given the context. They didn't have a model of truth; they had a model of what sounded plausible. Plausibility and accuracy were not the same thing.

"They hallucinate all the time," Altman acknowledged. "Well, not all the time. They used to hallucinate all the time. They now hallucinate a little bit."

For many use cases, this was acceptable. Brainstorming didn't require accuracy. Creative writing benefited from imagination. Coding assistance could be verified by running the code. The hallucinations were annoying but manageable.

For other use cases, hallucination was dangerous. Lawyers who trusted ChatGPT for legal research filed briefs citing nonexistent cases. Journalists who used it for fact-checking repeated fabricated claims. Medical information could be confidently wrong.

The limitations were real. They were also, to some extent, solvable. Each subsequent model improved on accuracy. Each update reduced the hallucination rate. The trajectory was clear even if the destination wasn't.

The Values Question

When Google finally rushed Gemini to market in early 2024, users discovered something strange. Ask it to generate an image of the founding fathers, and it produced a group of diverse women. Ask for a picture of the Pope, and it showed diverse women. The AI seemed to have been trained with such aggressive diversity requirements that it would override historical fact.

Elon Musk, who had co-founded OpenAI before departing and later founding xAI, called Demis Hassabis directly. "Why is Gemini lying to the public about historical events?" he demanded.

Hassabis's answer was revealing. His team at DeepMind hadn't programmed those behaviors. Another team at Google had added them after the fact, adjusting the model's outputs to meet corporate diversity goals.

"I do think that it matters how you build the AI and what kind of values you instill in the AI," Musk reflected. "My opinion on AI safety is the most important thing is that it'd be maximally truth seeking, that you don't force the AI to believe things that are false."

The incident illuminated a deeper problem. Language models were trained not just on internet data, but through human feedback: people telling the system which answers were "good" and which were "bad." Those feedback signals embedded values. And if those values prioritized certain outcomes over accuracy, the AI would learn to lie.

"The problem with that is that it can drive AI crazy," Musk argued. "Because it's trying to... you're telling AI to believe a lie. And that can have very disastrous consequences."

The Gemini controversy was embarrassing for Google. But it raised a question that would only become more pressing: who got to decide what values AI systems should hold? And what happened when those systems became more powerful than any previous technology?

For Musk, the answer was clear. After years of warning about AI risks from the sidelines, he had come to a different conclusion. "There's two choices here," he explained. "Either be a spectator or a participant. And if I'm a spectator, I can't really influence the direction of AI, but if I'm a participant, I can try to influence the direction of AI."

In July 2023, he founded xAI. The stated goal: build a "maximally truth-seeking" AI that was curious about the universe and didn't have ideological guardrails that made it unreliable.

The AI race now had another entrant. And the competition wasn't just about capability; it was about what AI should believe.

The Microsoft Investment

Microsoft moved differently.

In January 2023, Microsoft announced a $10 billion investment in OpenAI. The deal was complex: part capital, part partnership, part strategic positioning. Microsoft would integrate OpenAI's models into its products. Azure would provide the compute infrastructure. Both companies would benefit from the other's strengths.

The investment signaled something important about the AI landscape. Building frontier models required resources that only a handful of organizations could provide. The compute costs were enormous. Training runs required thousands of GPUs, costing tens of millions of dollars. The talent was scarce and expensive. The data pipeline was complex.

Microsoft wasn't just investing in OpenAI. Microsoft was investing in the GPU allocation that OpenAI would need to keep scaling. Access to compute had become a strategic asset, as important as the models themselves.

The partnership also solved a problem for OpenAI. The nonprofit structure had been awkward from the start. OpenAI needed massive capital to pursue its mission, but nonprofits couldn't easily raise venture funding. The capped-profit subsidiary structure, where investor returns were limited to 100x, was a compromise. Microsoft's investment provided resources without turning OpenAI into a conventional tech company.

Whether the compromise would hold was another question.

The Breadth of Impact

By February 2023, ChatGPT's influence was everywhere.

Students used it for essays. Programmers used it for code. Marketers used it for copy. Customer service teams used it for response templates. Lawyers used it for research. Doctors used it to explain diagnoses to patients. Writers used it for brainstorming. Researchers used it to summarize papers.

The use cases kept expanding. Each user discovered applications that the developers hadn't anticipated. The technology was genuinely general-purpose, a tool that could help with any task that involved language, which meant almost every task.

The implications were profound and unclear. If ChatGPT could write a serviceable essay, what happened to the millions of people whose job was writing serviceable essays? If it could generate competent code, what happened to entry-level programmers? If it could summarize documents and answer questions, what happened to analysts and researchers and consultants?

"You can say, 'Oh, man, it's doing everything I do. What's going to happen to me?'" Altman acknowledged. "Or you can say, like through every other technological revolution in history, 'Okay, now there's this new tool, I can do a lot more.'"

Both perspectives were valid. Both would prove correct, for different people, in different contexts.

The GPU Shortage

Everyone wanted what made ChatGPT possible.

Every technology company in the world was scrambling to build their own version. Microsoft was integrating AI into Bing and Office. Google was rushing to ship Bard. Meta was accelerating its research timeline. Startups were launching by the dozens.

They all needed the same thing: GPUs.

Training large language models required specialized hardware. Not just any chips, but specifically the kind of chips that could perform the parallel computations that neural networks demanded. The kind of chips that one company had spent seventeen years optimizing.

ChatGPT had done something no AI announcement had ever done. It made the technology real for people who had never thought about AI. The abstraction layer had collapsed. And now every company in the world wanted to build their own version.

There was just one problem.

There weren't enough GPUs.

The company that could provide them had been building toward this moment since 2006. Jensen Huang's bet on CUDA, the years of investment in parallel computing, the gaming market that had funded the R&D: all of it had been preparation for a moment nobody could have predicted.

The Renaissance had exploded into public consciousness. The builders had provided the foundations. Now came the race to scale.

INTERLUDE

THE LIBRARY THAT LEARNED TO READ

The story pauses here.

ChatGPT has just exploded. A hundred million users in two months. But what is it actually doing when it responds to you? The coffee shop taught machines to see. Now we need to understand how they read.

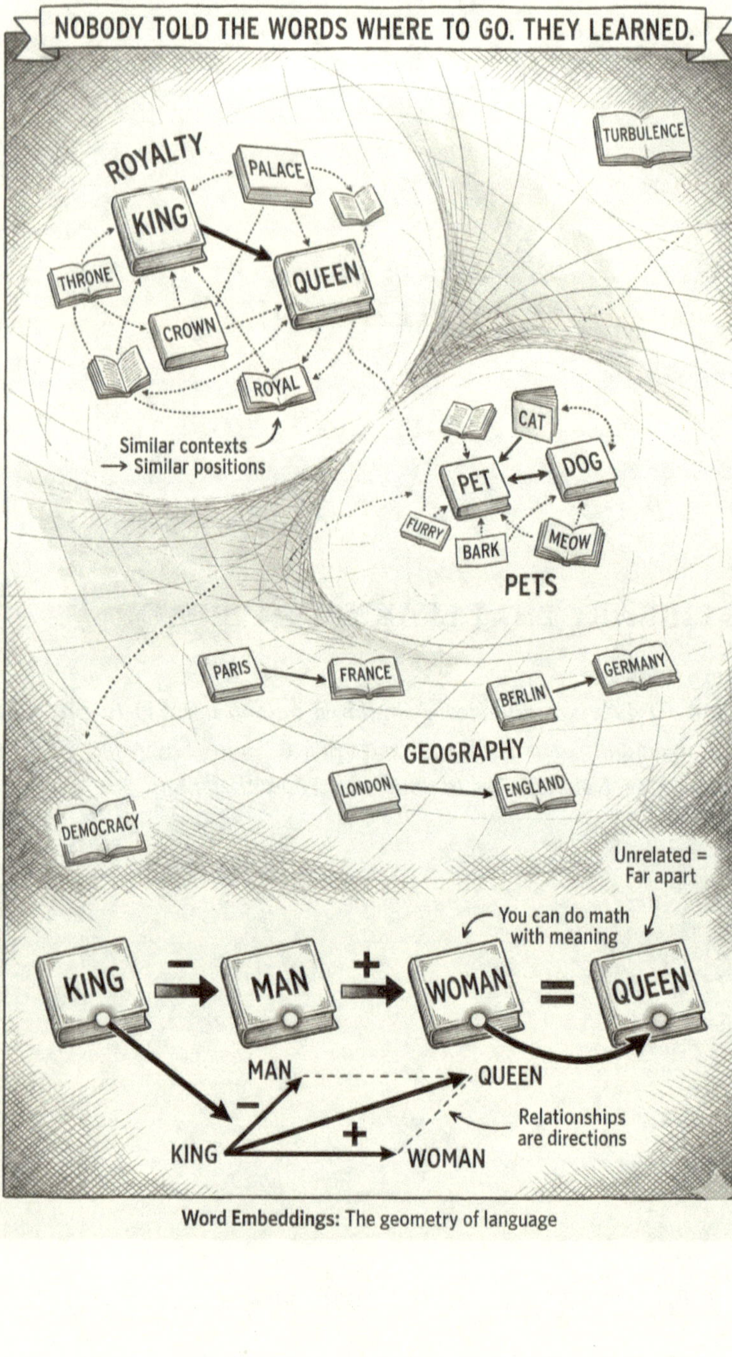

Word Embeddings: The geometry of language

In the first interlude, we taught a coffee shop to see. Workers learned to detect patterns – edges, shapes, objects – by passing blame backward through the chain.

But vision is only half the story. The other half is language.

In 2003, while Hinton was cracking deep networks and LeCun was teaching machines to read handwriting, Yoshua Bengio in Montreal asked a different question: How do you predict the next word in a sentence?

It sounds simple. "The cat sat on the ___" probably ends with "mat" or "floor" or "couch." Not "democracy." Not "turbulence." Your brain does this instantly, without effort. But how would you teach a machine?

The old approach was brute force. Count how often each word follows every other word in a giant pile of text. Store the counts in tables. "The" is followed by "cat" 0.02% of the time, by "dog" 0.01%, by "end" 0.03%...

The problem: language is combinatorial. There are hundreds of thousands of words. The number of possible two-word combinations is in the billions. Three-word sequences? Trillions. The tables explode. You can't store them. And even if you could, most sequences are so rare you'd never have enough examples to count.

Bengio needed a different approach. He found it in geometry.

The Self-Arranging Library

Imagine a library with a peculiar organization system.

In an ordinary library, books are shelved by category. History in one section, science in another, fiction across the hall. The physical location tells you something about the content; books near each other tend to be about similar things.

Now imagine a library where every word in the language has its own location. Not on a shelf, but in a vast high-dimensional space. The word "king" floats somewhere in this space. So does "queen." So does "throne" and "crown" and "palace."

But here's what makes this library magical: nobody told the books where to go. Nobody hand-coded that "king" and "queen" are related.

The books learned their positions by being read together, millions of times, in millions of sentences.

When "king" appears in similar contexts as "queen" ("The ___ ruled wisely," "The ___ wore a crown"), the library nudges them closer together. When words appear in different contexts, they drift apart.

Over millions of iterations, the library organizes itself. "Cat" settles beside "dog." "Paris" drifts toward "France," and both hover near "Berlin" and "Germany." "Sad" and "melancholy" and "grief" form their own little constellation.

This is what Bengio built. In AI terminology, each word's position is called an *embedding*: a location in meaning-space. Not three dimensions like a physical room, but hundreds of dimensions. Each dimension captures some aspect of meaning.

The magic: similar words have similar embeddings. And this similarity isn't programmed. It's learned.

The Arithmetic of Meaning

Here's something strange that emerged from this geometric view of language. You can do arithmetic with words.

Take the location of "king." Subtract "man." Add "woman." Where do you land?

Near "queen."

"Paris" minus "France" plus "Germany"? You land near "Berlin."

The library had discovered something about how meaning works. Relationships between words are encoded as directions in space. The direction from "man" to "woman" is the same as the direction from "king" to "queen." Gender is a direction. Capital-of is a direction. Past-tense is a direction.

Nobody programmed these relationships. They emerged from patterns in text. The library read enough sentences to discover the hidden geometry of meaning.

The Window Problem

Bengio's 2003 model had a limitation that would take a decade to overcome.

To predict the next word, the network looked at a fixed window of previous words. Five words back, perhaps. Maybe ten. The window slid along the sentence, making predictions, learning embeddings.

But meaning in language doesn't respect windows.

Consider: "The trophy wouldn't fit in the suitcase because it was too big."

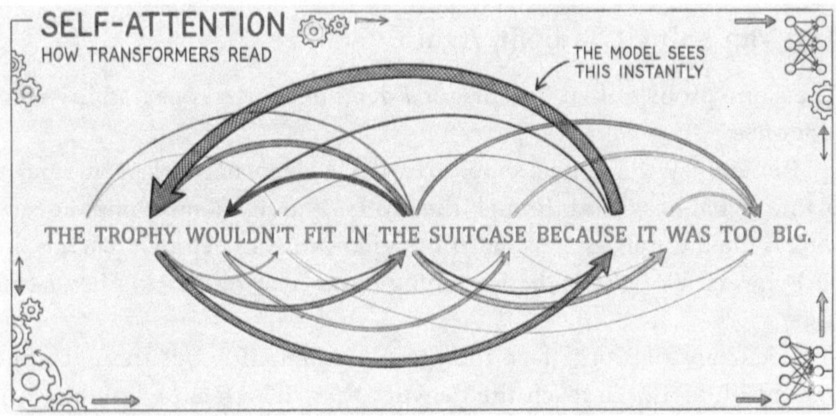

What does "it" refer to? The trophy. But to know that, you need to understand that trophies are typically bigger than suitcases, and that "too big" means the larger object is the problem. This isn't a local pattern. It requires world knowledge, reasoning, context that spans the entire sentence.

Now consider: "The author, who had written seventeen novels over a career spanning four decades and had won the Pulitzer Prize in 1987 for a work that critics initially dismissed but which later became required reading in graduate programs across the country, finally retired."

The subject and verb are separated by forty-three words. No fixed window captures this. The meaning stretches across the entire structure.

The community's solution was called the *recurrent neural network*, or RNN. Instead of looking through a fixed window, the network would process words one at a time, carrying forward a hidden state. A memory.

Think of the RNN as a librarian reading through a manuscript, word by word. She reads the first word, holds it in memory. Reads the second, updates her understanding. And so on, sequentially, through the entire text.

She remembers what she's read; she carries a mental summary that updates with each word. But she can never read two words at once. She can never jump ahead. She can never parallelize.

In theory, this solved everything. The memory could preserve information indefinitely, connecting "author" to "retired" across any distance.

In practice, the memory leaked.

The Vanishing Gradient, Again

The same problem that had plagued deep networks appeared in recurrent ones.

Remember the coffee shop. To train a deep network, you send a blame signal backward through the layers. The signal multiplies at each step. If the multipliers are small, the signal vanishes before it reaches the early layers. Workers at the beginning never hear feedback. They never learn.

Recurrent networks face the same mathematics, but through time instead of depth. To teach the network that "it" refers to "trophy," you need to send a learning signal backward through all the words in between. Each step multiplies. The signal vanishes.

Long-distance dependencies disappeared into mathematical noise. The memory forgot what it was supposed to remember.

Researchers developed clever architectures – Long Short-Term Memory networks (LSTMs) and Gated Recurrent Units (GRUs) – that mitigated the problem through gates and shortcuts. These helped. The models improved. Language modeling became a real research field, not just a backwater.

But the fundamental issue remained. Recurrent networks processed sequences step by step. Information had to survive the journey. And long journeys meant degraded signals.

And remember the hardware story? GPUs, the chips that powered video games, were brilliant at doing thousands of things in parallel. But they couldn't help an RNN. You can't parallelize sequential reading.

For a decade, the field assumed this was the price of processing language. Sequences were sequential. You had to walk through them one word at a time.

Then, in 2014, a student in Bengio's lab proposed something different.

The Reference Desk

Return to the library.

You're writing a translation. The source sentence is in French: "Le chat noir dort sur le tapis rouge." You need to produce English: "The black cat sleeps on the red mat."

The recurrent approach: read the entire French sentence, compress it into a single memory state, then generate English word by word. All the information about the source (the cat, its color, the mat, its color) must squeeze through a single bottleneck.

Think of it this way. A librarian reads an entire French sentence, writes a single summary card, hands the card to an English librarian who's never seen the original, and says: "Reconstruct the sentence from this."

For short sentences, it worked. But long sentences? The summary card couldn't hold everything. Details were lost. Nuance vanished. The translation came out mangled.

In 2014, Dzmitry Bahdanau proposed a fix.

Instead of compressing everything first, let the English librarian look back at the original French sentence while writing each word.

When she writes "cat," she glances back at the French text and focuses on "chat." When she writes "sat," she focuses on "assis." When she writes "on the mat," she attends to "sur le tapis."

Each output word gets to *attend* to different parts of the input. The network learns what to focus on.

This is *attention*: a learned mechanism for deciding what matters right now.

The librarian doesn't give equal weight to every French word when writing each English word. She focuses. She attends. And the focusing is learned, trained through millions of translation examples, adjusting weights with backpropagation, just like our coffee shop workers.

Suddenly, long sentences worked. The summary card was no longer a bottleneck. The network could handle nuance, ambiguity, complexity, because it could always look back at what it needed.

The Library Looks at Itself

Here is where the story accelerates.

Attention, as Bahdanau designed it, let a decoder look at an encoder. The English-generating system looked at the French source. Output looked at input.

But what if a sentence looked at *itself*?

Consider again: "The trophy wouldn't fit in the suitcase because it was too big."

What does "it" mean? To answer this, "it" needs to look at every other word in the sentence. It needs to compare itself to "trophy" and "suitcase" and evaluate which makes more sense given "too big."

This is *self-attention*. Every word looking at every other word to determine its own meaning.

Imagine the library rearranging itself in real-time. When you read "it," the word looks around at its neighbors (not just nearby neighbors, but every word in the sentence). It calculates: how relevant is "trophy" to my meaning? How relevant is "suitcase"? The weights determine how "it" should be understood.

This happens simultaneously for every word. "Trophy" is looking at "suitcase" and "fit" and "big." "Because" is looking at everything, calculating how it connects the clauses. The entire sentence is computing its own structure, in parallel.

Attention Is All You Need

In 2017, a team at Google asked a radical question: *What if we removed the sequential part entirely?*

What if, instead of one librarian reading word by word, every word in the sentence could see every other word simultaneously?

Not reading one at a time. Not carrying a running summary. Just... *all words, all at once, all attending to each other.*

They called it the transformer. And they titled their paper with a kind of audacity rare in academic writing:

"Attention Is All You Need."

No recurrence. No convolutions. No sequential processing. Just attention layers stacked on attention layers.

The architecture was almost absurdly simple. Take your input, a sequence of word embeddings. Pass it through a self-attention layer: every word looks at every other word, calculates relevance weights, and updates its representation based on the weighted average of all other representations.

Stack several of these layers. Add some feedforward networks in between. Add some normalization. That's it.

The magic was in what this enabled.

Recurrent networks were sequential. To process word fifty, you first had to process words one through forty-nine. This meant training was slow; you couldn't parallelize across positions.

Transformers were parallel. Every word could attend to every other word simultaneously. The computation could spread across thousands of GPU cores, which could finally unleash their full power on language.

The Scaling Discovery

What happened next surprised even the researchers.

The transformer was designed for translation. But it turned out to be a general-purpose architecture for processing sequences of any kind.

Text? Yes. Code? Yes. Music? Yes. Protein sequences? Yes. Anything that comes in order could be fed through the transformer's attention mechanism.

And then came the discovery that would reshape the field.

Bigger transformers were better transformers. More parameters. More data. More compute. The performance kept improving. Nobody found a ceiling.

OpenAI pushed the progression. GPT-1 in 2018: 117 million parameters. GPT-2 in 2019: 1.5 billion. GPT-3 in 2020: 175 billion. Each jump brought new capabilities that simply hadn't existed in smaller models.

This was the *scaling hypothesis*, already explored in the previous chapter: the idea that intelligence might be a function of scale. The library got bigger, the attention got sharper, and the machine got smarter.

The coffee shop learned to see. The library learned to read.

Vision came first: AlexNet in 2012, the ImageNet annihilation, the industry waking up to deep learning.

Language came second: embeddings, attention, transformers, the scaling hypothesis.

By 2020, the foundations were in place for something no one quite expected. A model trained simply to predict the next word, given enough parameters, enough data, enough compute, could write essays, answer questions, generate code, and carry on conversations.

On November 30, 2022, OpenAI would release that model to the public.

They called it ChatGPT.

And the world would never look at libraries, or language, the same way again.

A Translation Guide

For readers who encounter AI terminology elsewhere, here's how our library maps to the technical language:

Embedding A word's position in the library: its location in high-dimensional meaning-space. Similar words have similar embeddings. Relationships between words are encoded as directions in space.

Embedding Space The library itself: the high-dimensional room where words find their positions.

Language Model A system that predicts what comes next. Given "The cat sat on the," predict "mat" (or "floor" or "couch"). All the magic emerges from this simple objective, pursued at massive scale.

Token What the model actually sees. Not quite words. Common words are single tokens; rare words are split into pieces. "Understanding" might be "under" + "standing." The vocabulary is typically 50,000 to 100,000 tokens.

Attention The librarian looking back at specific books while writing. A learned mechanism for deciding what to focus on.

Self-Attention Every book looking at every other book in the same library: words attending to other words in the same sentence.

Recurrent Neural Network (RNN) The old approach: a librarian reading one book at a time, carrying a running summary. Sequential, slow, hard to parallelize.

LSTM (Long Short-Term Memory) A more sophisticated RNN with gates that control what to remember and what to forget. Helped with the vanishing gradient problem but never fully solved it.

Transformer The new architecture: every book sees every book simultaneously. No sequential bottleneck. Parallelizes beautifully on GPUs.

Encoder The part that reads the input: the French librarian processing the original text.

Decoder The part that generates output: the English librarian writing the translation, attending back to the original.

Context Window How many books the librarian can see at once. Early models: a few words. GPT-2: about a thousand tokens. GPT-4: over 100,000 tokens. Larger windows mean more context for understanding meaning.

Parameters The adjustable settings, like the coffee shop's weights. GPT-3 has 175 billion parameters. Each one was tuned through training.

Scaling Hypothesis The idea that bigger libraries with more parameters, trained on more data, will keep getting better. No ceiling in sight.

Large Language Model (LLM) A very big transformer trained on very much text. GPT-3, GPT-4, Claude, Gemini: all LLMs. The self-arranging library, scaled up.

Now you know how they read. The library that arranges itself is no longer magic – it's architecture you can understand.

INTERLUDE

THE TEACHER

How These Machines Actually Work

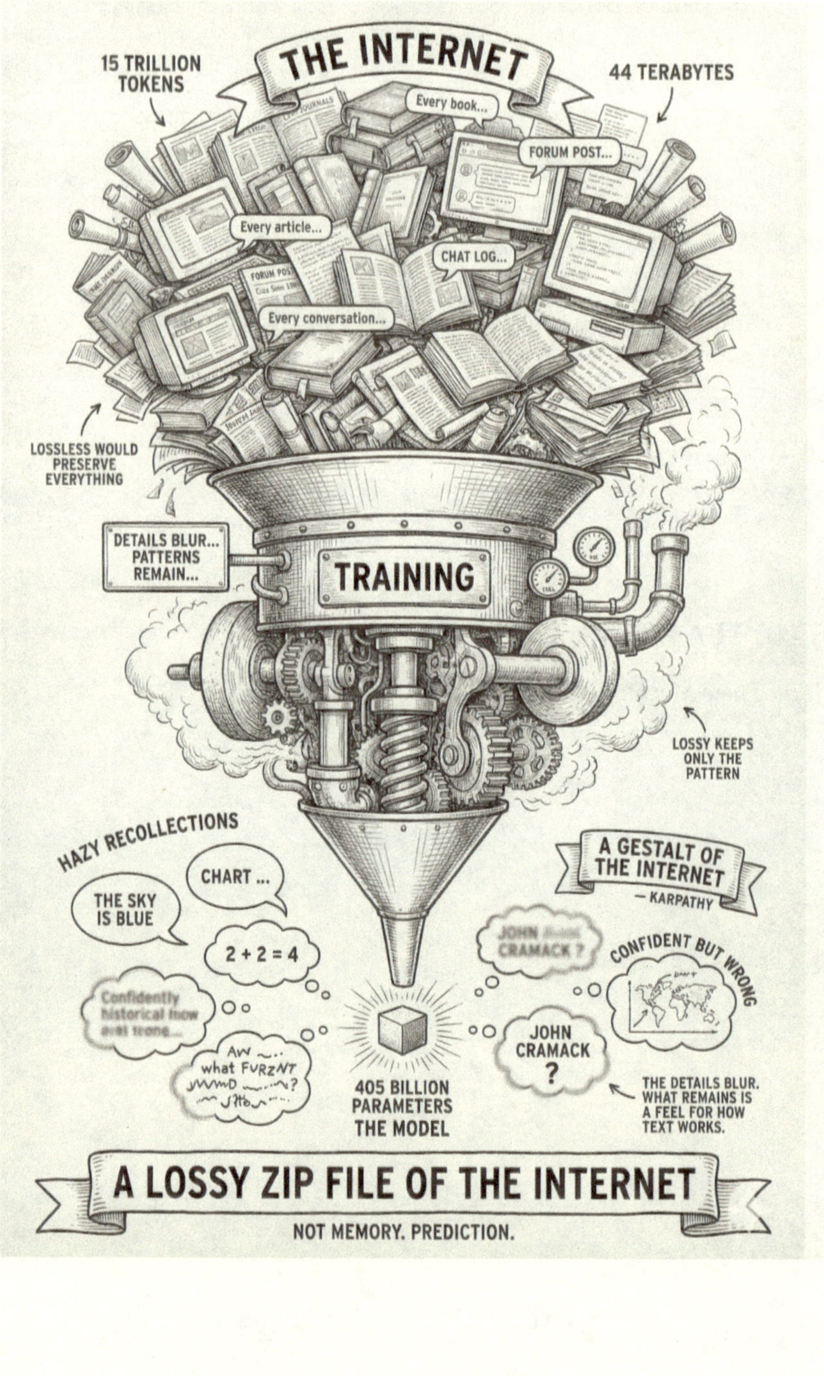
THE INTERNET
15 TRILLION TOKENS
44 TERABYTES
Every book...
FORUM POST...
Every article...
CHAT LOG...
Every conversation...
LOSSLESS WOULD PRESERVE EVERYTHING
DETAILS BLUR... PATTERNS REMAIN...
TRAINING
LOSSY KEEPS ONLY THE PATTERN
HAZY RECOLLECTIONS
CHART ...
THE SKY IS BLUE
2 + 2 = 4
A GESTALT OF THE INTERNET
— KARPATHY
CONFIDENT BUT WRONG
JOHN CRAMACK ?
405 BILLION PARAMETERS THE MODEL
THE DETAILS BLUR. WHAT REMAINS IS A FEEL FOR HOW TEXT WORKS.
A LOSSY ZIP FILE OF THE INTERNET
NOT MEMORY. PREDICTION.

The Counter

Two numbers were climbing in early 2023.

The first was ChatGPT's user count: the fastest adoption curve in technology history, crossing one hundred million in two months.

The second was quieter. On YouTube, a video called "Let's Build GPT from Scratch" was accumulating millions of views. Two hours long. No production budget. Just a man at a computer, building the thing that had captured the world's attention.

The man was Andrej Karpathy. He had helped build ChatGPT. He had led AI at Tesla. He had credentials that could command any position in the industry.

And he was giving away the secrets for free.

The Perpetual Student

"I stayed in school all the way until PhD," Karpathy says, "and then they wouldn't let me stay longer, so I went to industry."

He's not joking. Born in Bratislava, Slovakia, Karpathy discovered early that learning was the thing he loved most. Not learning *for* something; learning as the thing itself.

His path wound through the center of the AI revolution without him planning it that way. PhD at Stanford under Fei-Fei Li, who built ImageNet, the dataset that made deep learning's triumph possible. Founding member of OpenAI in 2015. Director of AI at Tesla, teaching cars to see.

At each stop, the same pattern emerged: Karpathy would master something, then figure out how to explain it to others. At Stanford, he created CS231n, the first deep learning class, which trained a generation. At Tesla, he gave talks that demystified neural networks for engineers. The teaching wasn't separate from the work. It *was* the work.

Then, in 2022, he made an unusual choice. He left Tesla. Returned briefly to OpenAI. And then left again to make YouTube videos.

The credentials said he should be running a lab. He chose to run a classroom instead.

What You're Actually Talking To

To use these systems well, you need to understand what they actually are. Karpathy has spent years developing the right mental models: ways of thinking about these tools that are neither dismissive nor mystical.

Start here: when you type something into ChatGPT and press enter, what happens?

The short answer is prediction. The system looks at everything you've written and predicts what words should come next. Then it generates those words, adds them to the conversation, and predicts again. And again. The whole response emerges from this simple loop: predict, append, predict, append.

"These are statistical token tumblers," Karpathy calls them. "It's just trying to sample the next token in the sequence."

Token tumbler. The phrase does work.

But this raises an obvious question: if it's just predicting words, why does it seem to *know* things? Why can it explain quantum mechanics, debug code, write sonnets?

The answer lies in how it learned to predict.

The Lossy Zip File

Here's the core insight, in Karpathy's words:

"You can think of these 405 billion parameters as kind of like a zip file. But it's not a lossless compression. It's a lossy compression. We're kind of left with a gestalt of the Internet."

A ZIP file preserves everything. Uncompress it, and you get the original back exactly.

Neural networks don't work that way. They compress 15 trillion tokens of training data, 44 terabytes of human text, into a much smaller space. Something has to give. The details blur. What remains is the *pattern*. The statistical skeleton of how humans write.

Think of it like this: if you read every book in a library, you wouldn't remember every sentence. But you'd develop an intuition for how sentences work, how ideas connect, how arguments flow. You'd have a *feel* for language that lets you generate plausible text about almost anything, even topics you can't recall precisely.

That's what these models have. A feel for how text works, extracted from everything humans have written.

"Anything that happens during training," Karpathy explains, "the knowledge is only kind of like a hazy recollection."

The things that appear often in the training data (common facts, typical patterns, standard explanations) are remembered well. The things that appear rarely get fuzzy. And made-up facts can feel just as real as true ones, because the model learned to produce *confident-sounding* text, not *accurate* text.

The Restaurant Kitchen

In Chapter 1, we met John Carmack and explained his approach to parallel processing through a restaurant kitchen analogy. A chef doesn't create ingredients from scratch; she combines, transforms, sequences. She has recipes (algorithms) and ingredients (data) and equipment (hardware), and her skill lies in orchestrating them efficiently.

Large language models work similarly, but with a twist.

The "ingredients" are patterns extracted from text, billions of them, compressed into parameters. The "recipe" is the transformer architecture, a mathematical structure that decides how to combine patterns given a particular input. The "equipment" is the same GPU hardware that Carmack's games helped create.

But here's the twist: the chef doesn't consciously know her recipes. She learned them by reading every cookbook ever written and developing intuitions about what goes with what. When you ask for a soufflé, she doesn't retrieve a recipe; she *feels* her way toward something soufflé-like, drawing on patterns she can't explicitly articulate.

This explains both the power and the limitation. The chef can improvise brilliantly, combining influences in novel ways. But she might also confidently serve you something that looks like a soufflé but isn't, because her intuition, trained on descriptions of soufflés, doesn't include the actual experience of baking one.

The DJ Set

How does the learning actually happen? Karpathy has an analogy for that too.

"Think of these parameters as kind of like knobs on a DJ set. As you're twiddling these knobs, you're getting different predictions for every possible input."

The model starts random. Billions of knobs, all set to meaningless positions. Feed it text, and it predicts garbage.

Then training begins. You show it a window of text, say a few thousand words from a web page, and ask it to predict what comes next. It guesses wrong. You measure how wrong. Then you adjust the knobs, just slightly, in the direction that would have made the guess less wrong.

Repeat this billions of times across billions of examples. The knobs gradually find settings that produce good predictions. The random noise becomes coherent text.

"Training a neural network just means discovering a setting of parameters that seems to be consistent with the statistics of the training set."

That's all. It's mechanical. The mechanism is vast, months of training on thousands of GPUs, billions of dollars in compute, but the principle is simple. Find knob settings that match the patterns in the data.

Working Memory

"Anything in the weights is like a hazy recollection of what you read a year ago. Anything you give it as context is directly in working memory."

This insight unlocks everything about how to use these systems effectively.

The model's *parameters*, the billions of numbers from training, are long-term memory. Fuzzy. Compressed. Good for common patterns, bad for rare details.

The *context window*, the text you provide when you ask a question, is working memory. Crystal clear. Directly accessible. The model can see it like notes on a desk.

This is why prompting matters. When you paste a document into the chat before asking questions about it, you're not trying to jog the model's memory. You're putting information directly into its working memory, where it can use it without distortion.

The model doesn't need to "recall" what's in the context window. It can see it. The information is right there, as accessible as the words on a page in front of you.

This also explains a practical tip: if you want accurate information about something specific, don't ask the model to remember. Give it the source material and ask it to work with that.

Ghosts, Not Animals

Here's perhaps Karpathy's most important mental model:

"We're not building animals. We're building ghosts."

The distinction cuts deep.

Animals come from evolution. A zebra stands minutes after birth and follows its mother, not because it learned to, but because millions of years of optimization encoded that behavior into its DNA. The hardware comes pre-wired.

AI systems come from training data. They read the Internet and learn to predict what word comes next. No evolution. No embodiment. No instincts written into silicon.

"Sort of ethereal spirit entities," Karpathy calls them. "Fully digital, kind of mimicking humans. A different kind of intelligence."

Ghosts that know everything humans have written but have never held a cup or felt cold. Ghosts that can write poetry about heartbreak without having a heart. Ghosts that can explain how to ride a bicycle without ever having a body.

The metaphor matters because it sets expectations correctly. These aren't minds like ours. They're something new: powerful and strange and worth understanding on their own terms.

When you talk to ChatGPT, you're not talking to a digital human. You're talking to a ghost, an entity that exists entirely as patterns in numbers, that has no continuous existence between conversations, that learned everything it knows from human text but has never experienced anything humans experience.

Understanding this changes how you interpret its outputs. The confident tone doesn't mean certainty. The helpful demeanor doesn't mean desire to help. These are patterns learned from text, not experiences felt by a mind.

Why Hallucinations Happen

With these mental models in place, hallucinations make sense.

Karpathy demonstrates with a fake name. Ask an older model "Who is Orson Kovats?" and it confidently invents: an American author, a 1950s TV character, a baseball player. Different answer each time, same confident tone.

Why? Because of how the training data looked.

In the training set, questions of the form "Who is [famous person]?" are always answered confidently. The human labelers who created the training examples knew who they were writing about; they looked things up, wrote accurate answers, used a confident tone.

The model learned the *shape* of answers without learning to check if they're true.

"Even if some part of the network kind of knows that it doesn't know," Karpathy explains, "saying 'I don't know' is not going to happen because the model statistically imitates its training set."

The model has never seen examples where the correct response is uncertainty. So it does its best, which means guessing confidently in the style it learned.

Modern systems have improved through a technique Karpathy describes: interrogate the model to find the boundaries of its knowledge, then add training examples where the correct answer is "I don't know." This teaches the model that uncertainty is sometimes the right response.

But the fundamental issue remains. These are pattern-matching systems. They produce text that *looks* right, that matches the statistical patterns they learned. Whether that text is *actually* right is a separate question, one the model cannot directly answer.

Swiss Cheese Capabilities

"I call it the Swiss cheese model of capabilities," Karpathy says. "There are holes in the cheese."

The model can solve complex calculus problems but stumble on simple arithmetic. It can write elegant code but fail to count the letters in a word. It can explain physics at a graduate level but not tell you that 9.11 is smaller than 9.9.

Why? Because these systems don't have unified understanding. They have patterns, millions of them, overlapping and interacting. Some patterns are strong (common in training data, well-reinforced). Others are weak or missing entirely.

The holes aren't random. They follow from the training process. Things that appeared often in training, in contexts where correctness mattered, tend to work well. Things that appeared rarely, or in contexts where plausibility was enough, tend to fail.

Mental arithmetic fails because training data rarely shows step-by-step arithmetic; humans just write the answer. Letter counting fails because text doesn't represent words as sequences of characters. To the model, "strawberry" is just a token, not s-t-r-a-w-b-e-r-r-y.

Understanding the Swiss cheese model changes how you use these tools. You check their work. You don't assume competence in one area implies competence in another. You learn where *your* model's holes tend to be, and you verify when the stakes matter.

The Simulation

Put all this together and you get a clearer picture of what you're interacting with.

When you use ChatGPT, you're not accessing intelligence in any deep sense. You're running a simulation: a very sophisticated statistical simulation of what a helpful, knowledgeable human might write in response to your prompt.

"It's a lossy simulation of a human that is kind of restricted in certain ways," Karpathy explains. The simulation is based on patterns extracted from human text. It produces outputs that match human patterns. But it's not a human, and the gaps between simulation and reality can matter.

The simulation doesn't have experiences. It doesn't remember previous conversations (each chat starts fresh). It doesn't have desires or preferences or goals, though it learned to write as if it does, because humans who write text often express desires and preferences and goals.

This framing is useful precisely because it's deflationary. It strips away the mysticism. You're not talking to a new form of consciousness. You're running a very good pattern-matching engine, one that can do impressive things within its domain and nothing at all outside it.

The Thinking Models

Something new is emerging, though. Karpathy is careful to distinguish it.

The models most people use, GPT-4, Claude, Gemini, are primarily products of the training process we've described. They predict text. They're impressive but fundamentally interpolative, drawing on patterns they've seen.

The newer "thinking" models add something different: reinforcement learning on reasoning tasks.

"These thinking models do use RL," Karpathy explains. "They go through this third stage of perfecting their thinking process and discovering new thinking strategies."

The result is systems that can, in some domains, do something more like genuine problem-solving: exploring possibilities, backtracking, finding solutions that aren't simple interpolations from training data.

"In principle, this paradigm is capable of doing something really cool, new, and exciting – something that no human has thought of before."

This is where Karpathy sees real novelty. Not in the base models, which are (very sophisticated) imitation engines. But in the RL-trained reasoning models, which are starting to discover strategies of their own.

The parallel to AlphaGo's Move 37 is explicit. AlphaGo found a move no human had ever played, through reinforcement learning on the game of Go. The thinking models might be capable of analogous discoveries in open-ended reasoning.

"These are primordial models for now," Karpathy cautions. The capability is nascent. But the direction is visible.

The Parallel

John Carmack released his source code after every game shipped. The engine that powered *Doom*, the tricks that made *Quake* possible, all eventually public. Anyone could read it. Anyone could learn.

Carmack believed opacity was the enemy of progress. If people couldn't see how things worked, they couldn't make them better.

Karpathy believes the same thing, applied to a different domain. Carmack showed how the graphics worked. Karpathy shows how the mind works, or at least how these new artificial minds work.

Both are acts of faith in other people. Faith that if you give away the knowledge, more will come back than you lost.

Eureka

Karpathy could be running an AI lab. He's building a school instead.

"Eureka is trying to build the Starfleet Academy."

From *Star Trek*. The institution where cadets learn to navigate the future.

"Education is the technical process of building ramps to knowledge. Lots of eurekas per second – that's what I want."

Ramps, not walls. Understanding per second as the metric. The goal isn't to gatekeep. The goal is to get as many people as possible to the insight as fast as possible.

He learned Korean recently. First alone, from the Internet: slow and frustrating. Then in a class, but the teacher had to serve everyone. Then with a private tutor.

"She instantly understood where I was as a student. I felt like I was the only constraint. I was always given the perfect information."

That's the target. An AI that can do what the tutor did: diagnose exactly where you are, serve exactly what you need, never too hard, never too easy.

"No LLM does that today, not even close. But I think it's tractable. Someone should build it."

He's trying to be that someone.

The Tools Arrive

The technology had been built. Jensen's chips. Carmack's demand. Hinton's ideas.

The product had been shipped. ChatGPT. Claude. Gemini. A hundred million users in months.

Millions watched Karpathy's videos. Some became researchers. Some became engineers. Some just became better users, people who understood, finally, what they were talking to.

"I actually really do care about humanity. Everyone has to be superhuman in a certain sense."

Not superhuman like a comic book. Superhuman like augmented. Amplified. Capable of things that unaugmented humans can't do.

The tools are arriving. The question is whether people will understand them well enough to use them wisely.

The videos keep accumulating views. Somewhere right now, someone is watching "Let's Build GPT from Scratch" for the first time, pausing, typing, running the code, watching it work.

They're learning what these ghost intelligences actually are. They're developing intuitions for when to trust and when to verify. They're preparing for a world where these tools are everywhere.

That's the gift that keeps giving.

Now you know how they think. The ghost is no longer unknowable – and you can decide what to build with it.

ACT III: THE SCALING FAITH

THE RACE (2023–2024)

ChatGPT proved the cathedral matters. Now everyone races to control it. The scaling laws become an article of faith.

OPENAI
ANTHROPIC
GOOGLE
XAI
META
loss curve
Loss
loss curve
SCALING LAWS:
CAPABILITY GROWS
WITH SCALE

13

THE KINGMAKER

The Phone Calls

Early 2023. Jensen Huang's phone rang constantly.

CEOs of the world's largest companies, begging for chips. Microsoft, Google, Meta, Amazon, Oracle: everyone wanted Nvidia's H100 GPUs. The chips were the hardware required to train and run large language models. Without them, you couldn't compete in AI.

Wait times stretched to months. A year in some cases. Prices soared. The H100 cost between $30,000 and $70,000 depending on configuration, when you could get one at all. Companies hoarded chips like strategic reserves. Nations implemented export controls. GPU allocation became a geopolitical issue.

The infrastructure Jensen had spent seventeen years building was suddenly the foundation for everything.

"All day long, I call people and beg them to give us their GPUs," Altman (OpenAI) admitted in interviews. "We are so incredibly constrained."

The situation was almost absurd. OpenAI had created the most viral product in technology history, reaching a hundred million users in two months. And their growth was bottlenecked by hardware supply. They had more demand than they could possibly serve.

The bottleneck had a single point of control: Nvidia.

The Bet That Took Seventeen Years

To understand why Nvidia dominated, you had to go back to 2006.

CUDA. The programming platform that let developers write code for Nvidia's graphics cards. At the time, GPUs were for games. Nobody used them for general-purpose computing. The idea that graphics chips could do useful work beyond rendering pixels was speculative at best.

Jensen Huang committed over a billion dollars to building out CUDA anyway.

"We bet the company on CUDA," Jensen recalled. "It was over a billion dollars... If we had waited for the market to tell us what to do, we would have been too late."

The reaction from the market was brutal. "When I launched CUDA, the audience was complete silence," Jensen remembered. "No customer wanted it, nobody asked for it, nobody understood it. Nvidia was a public company." The stock cratered. "Our valuation went down to like two billion or three billion, from about twelve billion. I crushed it in a very bad way."

The bet took years to validate. For a long time, CUDA was a solution in search of a problem. Scientific computing adopted it gradually. Cryptocurrency mining provided unexpected demand. But the killer application remained elusive.

AlexNet launched the revolution. In 2012, when Alex Krizhevsky (Toronto→AlexNet paper) and Ilya Sutskever (OpenAI→Safe Superintelligence) trained their image recognition network on two Nvidia GTX 580 graphics cards, they demonstrated that neural networks could run efficiently on GPU hardware. The deep learning revolution had begun, and it ran on Nvidia's chips.

"The GTX 580 SLI was the revolutionary computer that put deep learning on the map," Jensen explained. "We were lucky because we were inventing this technology, this computing approach. We were lucky that they found it."

Lucky. But also prepared. The CUDA ecosystem (thousands of software libraries, years of optimization, developer education, enterprise relationships) meant that when AI researchers needed to scale their experiments, Nvidia was ready. AMD could design comparable hardware. Intel could invest in AI chips. What they couldn't replicate was seventeen years.

And now? "It's higher," Jensen said of the stock price, allowing himself a rare moment of understatement about a company worth trillions. "But it changed the world. That invention changed the world."

The Moat

By 2023, Nvidia's position was nearly unassailable.

The H100 wasn't just a faster chip. It was the center of an ecosystem. CUDA had become the standard programming interface for AI development. Every major deep learning framework (PyTorch, TensorFlow) was optimized for Nvidia hardware. Researchers wrote their code assuming Nvidia chips. Switching to a competitor meant rewriting years of accumulated work.

This was Jensen's real insight. You didn't win in technology by building better hardware. You won by building platforms that became indispensable. Microsoft understood this with Windows. Apple understood this with iOS. Jensen understood it with CUDA.

"The thing to keep in mind," Jensen explained, "is what is the market opportunity that you're engaging? And that informs your size."

He had been told, years earlier, that Nvidia could never be larger than a billion dollars. It was a chip company, after all. Chip companies had natural ceilings.

But Nvidia wasn't just a chip company. It was the infrastructure provider for the most important technology transition in decades. They had built an entire ecosystem: working with game developers, getting games ported and adapted to their silicon, turning what started as a technology business into a platform business.

The same playbook that worked for games now worked for AI. The platform wasn't the chip. The platform was everything built on top of it.

The moat ran deeper than most realized. Sebastian Raschka, a machine learning researcher at Lightning AI and author of *Build a Large Language Model From Scratch*, had seen it firsthand. "The moat of Nvidia is probably not just the GPU," he reflected in early 2026. "It's more like the CUDA ecosystem, and that has evolved over two decades. Even back when I was a grad student, I was in a lab doing biophysical simulations, molecular dynamics, and we had a Tesla GPU back then. That was fifteen years ago. They built this up for a long time, and that's the moat."

Fifteen years. Before AlexNet. Before deep learning. Before anyone knew AI would matter. Jensen had been planting seeds in research labs that wouldn't bear fruit for a generation.

The Vertical Climb

The numbers were staggering.

In January 2023, Nvidia was worth approximately $360 billion. A large company, certainly, among the most valuable semiconductor firms in history. But not unusual by tech giant standards. Apple was worth $2 trillion. Microsoft was worth $1.8 trillion.

By June 2023, Nvidia crossed $1 trillion in market capitalization. It was the first chipmaker ever to reach that milestone.

By the end of 2023, Nvidia exceeded $1.5 trillion.

The velocity was unlike anything in corporate history. The company was adding hundreds of billions of dollars in value in months, not years. Each earnings report showed accelerating revenue. Each conference call discussed supply constraints, not demand concerns. The problem wasn't selling chips. It was making enough of them.

And yet Jensen's psychology hadn't changed. Still thirty days from going out of business. Still waking up afraid.

"But you guys are one of the biggest companies on planet earth," Rogan pushed back.

"The feeling doesn't change. The sense of vulnerability, the sense of uncertainty, the sense of insecurity: it doesn't leave you."

The psychology of scarcity, forged through decades of near-death experiences, meeting unlimited success.

The GPU Rich and the GPU Poor

The AI industry bifurcated.

A handful of organizations (OpenAI, Google DeepMind, Meta AI, Anthropic) had accumulated significant GPU capacity. They could train frontier models. They could run experiments at scale. They could push the boundaries of what was possible.

Everyone else was "GPU poor." Startups trying to compete, university researchers trying to publish, corporations trying to deploy, all of them constrained by hardware they couldn't obtain.

The distribution created a strange new hierarchy. Talent mattered, but compute mattered more. A brilliant researcher at a GPU-poor organization could be outperformed by a competent team with access to ten thousand H100s. The scaling laws that Ilya Sutskever had championed were becoming self-fulfilling: if bigger was reliably better, then whoever could train bigger would win.

This was the logic that drove the arms race. Not just between companies, but between nations. Access to advanced GPUs became a matter of national security. Export controls proliferated. The semiconductor supply chain (concentrated in Taiwan, dependent on Dutch lithography machines, optimized for global trade) suddenly looked fragile.

Jensen found himself at the center of geopolitics. Nvidia chips were essential for AI development. AI development was essential for economic competitiveness. Economic competitiveness was essential for national power. The logic chain ran straight from a graphics card company to the balance of global influence.

"The different businesses that Nvidia engages have such different natural sizes," Jensen observed. "But if you have the ability to see the opportunity and you are persistent enough and you're willing to dedicate the time, I think this opportunity is open to everybody."

The opportunity was open. The question was who could afford the entry fee.

The Singular Force

There was a deeper question: Would any of this have happened without Jensen?

"Isn't it a real case to be made," Lex Fridman asked in early 2026, "that without Jensen there's not a reinvigoration of the deep learning revolution? It could have been twenty years later."

Nathan Lambert, post-training lead at the Allen Institute for AI, didn't dismiss it. "The GPUs would still exist at the time of AlexNet and at the time of the Transformer. It was just hard to know if it would be one company as successful or multiple smaller companies with worse chips." He paused. "I just can't see Intel or AMD doing what Nvidia did. I don't think it would be a company that exists."

The comparison that kept surfacing was Steve Jobs.

"Everyone says the company is super-oriented around Jensen and how operationally plugged in he is," Lambert noted. "It sounds so unlike many other big companies. So long as that's the culture, I think I will expect them to keep progress happening. He's still in the Steve Jobs era of Apple."

What's Nvidia without Jensen? The question hung over every conversation about the company's future.

"People make things earlier and faster," Raschka noted. "Many great scientists credit being in the right place at the right time. In that way, Jensen is helping manifest this GPU revolution much faster and much more focused than without having a person there."

The counterfactual was impossible to prove. But the industry increasingly believed it: Jensen Huang hadn't just ridden the AI wave. He had made it possible.

The Hyperscaler Response

The hyperscalers were not passive.

Google, Microsoft, and Amazon (the three companies that controlled the cloud) each responded to the AI moment differently. But all three recognized the same thing: whoever controlled the infrastructure would shape what got built.

Google had the deepest AI roots. DeepMind, acquired in 2014, had produced AlphaGo and AlphaFold. Google Brain had invented the transformer. The company had more AI research talent than anyone. But Google also had the most to lose: its search monopoly, built over two decades, was suddenly vulnerable to conversational AI.

"The thing that is actually really exciting about this moment in AI," Sundar Pichai (Google) said carefully, "is how broad it is. It's not just one thing."

The breadth was real. AI was transforming search, advertising, cloud services, productivity software, consumer devices: every business Google touched. The company responded by merging Brain and DeepMind, rushing Gemini to market, and integrating AI across its product line. The strategy was defensive as much as offensive. If AI was going to disrupt search, Google wanted to be the one doing the disrupting.

Microsoft moved faster.

Satya Nadella (Microsoft) had been positioning for this moment since becoming CEO in 2014. The partnership with OpenAI, announced in 2019 and expanded repeatedly, gave Microsoft access to the most advanced language models in the world. Azure provided the compute. Office provided the distribution. GitHub provided the developer relationship.

"We're going to infuse AI into every layer of the tech stack," Nadella declared. Copilot for Microsoft 365. Copilot for GitHub. Copilot for Windows. The word "copilot" appeared everywhere, a branding strategy that positioned AI as assistant rather than replacement.

Amazon played a different game.

AWS had always been about selection and price. Amazon didn't need to build the best AI; it needed to ensure that every AI ran on Amazon's cloud. The company invested $4 billion in Anthropic, not for exclusive access but for a preferred partnership. It developed its own Bedrock platform, offering multiple foundation models from multiple providers. The strategy was classic Amazon: let others compete on features while Amazon competed on infrastructure.

"At AWS, our culture is to figure out how to be customer-obsessed," Andy Jassy (Amazon) explained, "and so we try to work backwards from what we think customers want."

What customers wanted was choice. And Amazon was happy to provide it, as long as the choices all ran on AWS.

Breaking the Dependency

All three hyperscalers were investing in alternatives to Nvidia.

The motivations were obvious. Nvidia's data center revenue had grown from $3 billion in 2020 to over $50 billion by 2024. By December 2025, the company's market cap had crossed $4 trillion. Every dollar flowing to Nvidia was a dollar not captured by the hyperscalers themselves. And the concentration of power created strategic vulnerability: a single supplier controlling the essential hardware for the AI revolution.

Amazon went furthest. Its Trainium chips, developed by Annapurna Labs (acquired in 2015), were designed to challenge Nvidia's dominance directly. "We've built a chip there, our own custom AI chip called Trainium," Jassy explained, "that's going to help people save a lot of money relative to what the cost has been to date."

Anthropic's expanded partnership explicitly committed to using Trainium for training future Claude models. "We're writing low-level kernels that allow us to directly interface with the Trainium silicon," Anthropic explained. "Our engineers work closely with Annapurna's chip design team to extract maximum computational efficiency from the hardware, which we plan to leverage to train our most advanced foundation models."

The bet was long-term. Nvidia's CUDA ecosystem (the same bet Jensen had made in 2006) created switching costs that would take years to overcome. Every AI researcher had trained on Nvidia hardware. Every framework was optimized for CUDA. Every piece of production code assumed Nvidia chips. Switching meant rewriting everything.

Microsoft was developing custom chips through partnerships. Google had TPUs, and the advantage of vertical integration, since their chips powered their own research. All three were patient. They had to be.

The irony was not lost on anyone: Jensen's bet on building an ecosystem, made when nobody cared about GPU computing, had created a moat so deep that trillion-dollar companies were spending billions to dig around it. The moat wasn't the hardware. The moat was seventeen years of accumulated software.

Different Optimizations

Step back and the pattern becomes clear.

Google optimizes for research excellence and distribution. It has the best scientists, the most ambitious moonshots, and the largest reach. Its weakness is product execution: turning breakthroughs into shipped software. But when Google finally ships, billions of people are already waiting.

Microsoft optimizes for enterprise integration. Its AI isn't just a chatbot. It's embedded in Office, Windows, Azure, GitHub. The value comes from AI that works where work already happens. "The scaffolding is what enables you to win," Nadella said.

Amazon optimizes for platform economics. It doesn't need to win the model race; it needs to ensure the models run on AWS. Selection plus low prices, applied to intelligence. Let a thousand models bloom, as long as they bloom in Amazon's cloud.

None of them need AI to be their primary business. They need AI to strengthen their existing positions. This makes them both powerful and constrained. Powerful because they have resources that startups can't match. Constrained because they have existing businesses to protect.

The frontier labs (OpenAI, Anthropic, and others) are racing to build AGI. The hyperscalers are racing to make sure AGI runs on their infrastructure.

The Jensen Return

The Nvidia CEO had been present throughout this book without being its protagonist.

Act I told his origin story: the immigrant child who cleaned toilets in rural Kentucky, the near-death experiences of NV1 and the Sega lifeline, the CEO who built Nvidia from three people in a Denny's to a corporation that nearly died multiple times. The leather jacket. The company meetings where he stood because his back hurt too much to sit.

The CUDA bet in 2006 was an act of faith. Jensen believed that GPUs could be used for more than graphics. He invested when nobody else cared. He built an ecosystem when the market didn't exist.

Building Nvidia turned out to be a million times harder than any of them expected. If they had realized the pain and suffering, the vulnerability, the embarrassment and shame – nobody in their right mind would have started. Jensen has said as much.

And yet here he was. Running the company everyone else depended on. Watching seventeen years of patient investment pay off beyond any reasonable expectation.

"I trick my brain into thinking, 'How hard can it be?'" Jensen explained. "Because you have to."

Trillions Mobilized

By 2025, the hyperscalers had committed trillions of dollars to artificial intelligence.

Google had merged its AI organizations, recovered from early stumbles, and integrated Gemini across its ecosystem. Microsoft had leveraged its OpenAI partnership into the most valuable company in the world while quietly building its own capabilities. Amazon had positioned itself as the neutral platform, investing in Anthropic while developing its own chips and models.

The infrastructure was built. The bets were placed. The strategies were deployed.

But beneath all the strategic differentiation lay a shared conviction. Google, Microsoft, Amazon, OpenAI, Anthropic, Meta: they disagreed on many things. Open versus closed. Safety versus speed. Consumer versus enterprise. But on one question, they were unanimous.

Bigger was better.

The Trillion-Dollar Question

The question everyone wanted answered: Was this sustainable?

Nvidia's dominance was real. The CUDA ecosystem was entrenched. The H100 was the gold standard for AI training. But technology markets were brutal. IBM had dominated computing until it didn't. Nokia had dominated mobile until it didn't. There was no guarantee that today's moat would hold tomorrow.

The competitors were coming. AMD was shipping its MI300 accelerators, claiming better efficiency for some workloads. Intel was developing Gaudi AI processors at lower price points. Google was expanding its TPU program. Amazon was building Trainium chips for AWS. Every cloud provider had reasons to reduce dependence on Nvidia.

And then there was China. Export controls limited Nvidia's sales to Chinese customers. The Chinese semiconductor industry was investing heavily in domestic alternatives. The geopolitical dimensions of GPU supply were becoming increasingly complex.

"The question is," Jensen observed, "what is the purpose of a job? What is the purpose of our company?"

His answer was characteristically ambitious. Nvidia wasn't just selling chips. Nvidia was building the infrastructure for intelligence itself. The market wasn't limited by the number of servers or data centers or even computing applications. The market was limited only by humanity's appetite for intelligence.

And that appetite, it turned out, was unlimited.

The Cathedral's Foundation

In October 2012, two graphics cards trained a neural network that changed everything.

Those were Nvidia GPUs. They ran software that Alex Krizhevsky had written using CUDA. They connected to a monitor through interfaces that John Carmack's engines had helped popularize. They embodied theories that Geoffrey Hinton had developed over forty years.

None of the people who made the AI revolution planned to make it together. Carmack wanted better games. Jensen wanted graphics chips to matter. Hinton wanted to understand the brain. Three obsessions. One accidental convergence.

By late 2023, the cathedral was visible. The infrastructure Jensen had built. The theory Hinton had developed. The demand Carmack had created. The scaling that Ilya believed. The product that Sam shipped.

ChatGPT was the interface. Nvidia was the infrastructure. The revolution was underway.

The foundation was laid. The platform was built. Trillions of dollars were mobilizing. And everyone (the hyperscalers, the frontier labs, the investors, the researchers) shared one conviction:

Scale was the answer.

Train bigger models. Buy more GPUs. Raise more capital. The loss curves didn't bend. The capabilities kept emerging. The scaling laws held.

It had become an article of faith.

14

THE FAITH

The Manifesto

The essay was only 1,116 words. It explained why seventy years of AI research had been wrong.

Richard Sutton (Reinforcement Learning) published "The Bitter Lesson" on his personal website in March 2019. He was sixty-two, a professor in Edmonton, Alberta, not famous outside of AI research. He had spent his career in Canada, far from Silicon Valley, developing the mathematical foundations for how machines learn from experience. His textbook, *Reinforcement Learning: An Introduction*, had quietly shaped a generation of researchers. He would share the Turing Award with Andrew Barto in 2024. But in 2019, he was simply a respected theorist who had noticed something important about his field.

The lesson was this: for seventy years, AI researchers had tried to build intelligence by encoding human knowledge into their systems. They had handcrafted features, designed expert rules, built symbolic representations of how they thought thinking worked. And for seventy years, these approaches had been superseded by simpler methods that just used more computation.

"The biggest lesson that can be read from 70 years of AI research,"

Sutton wrote, "is that general methods that leverage computation are ultimately the most effective, and by a large margin."

The two methods that worked were learning and search. Not human expertise. Not careful engineering. Just algorithms that could scale with more compute.

This was bitter, Sutton argued, because it meant that all the clever work researchers did (all their insights into how problems should be solved) was ultimately less valuable than simply throwing more computation at the problem. "We have to learn the bitter lesson that building in how we think we think does not work in the long run."

The essay spread through the AI community like a creed. It gave theoretical justification to what the practitioners already suspected: that the winning strategy was not cleverness but scale. More parameters. More data. More compute.

The age of faith had begun.

The Recipe

Ilya Sutskever (OpenAI→Safe Superintelligence) understood what Sutton's essay meant before most people had read it.

By 2019, Sutskever had already bet his career on the scaling hypothesis. He had left Google for OpenAI in 2015, trading the security of the most profitable company in the world for a nonprofit startup with no products and a mission that sounded like science fiction. He had co-authored GPT-2, which could generate coherent paragraphs of text, a capability that seemed miraculous at the time.

Now, reflecting on what had happened, he saw the deeper pattern.

"The way ML used to work is that people would just think of stuff and try to get interesting results," he explained. "Then the scaling insight arrived. Scaling laws, GPT-3. And suddenly everyone realized we should scale."

He paused on the word itself.

"Scaling is just one word, but it's such a powerful word because it informs people what to do. They say, okay, let's try to scale things."

The power of the scaling insight was not just intellectual. It was operational. Before scaling, AI research was unpredictable. You funded researchers, they tried things, sometimes they worked, usually they didn't. The field advanced through lucky breakthroughs that no one could schedule.

Scaling changed that calculus.

"Companies love this because it gives you a very low-risk way of investing your resources," Sutskever observed. "It's much harder to invest your resources in research. Compare that. If you research, you need to go forth to researchers and research and come up with something versus get more data, get more compute. You know you'll get something from pre-training."

This was the key. *You know you'll get something.* The scaling laws, first articulated by Jared Kaplan and Dario Amodei (both OpenAI→Anthropic) and colleagues in 2020, showed that loss decreased predictably as you increased model size, data, and compute. There was a power law relationship. Double the parameters, the model gets better by a predictable amount. Double the data, same thing. Double the compute, same thing.

For the first time in AI history, progress was something you could *buy*.

The Believers

The scaling faith spread fast.

At OpenAI, Sam Altman was raising billions of dollars to test its limits. The logic was simple: if more scale produced more capability, then the path to artificial general intelligence was a financing problem. Raise enough money. Buy enough GPUs. Train a big enough model. The math would take care of the rest.

"I think this is unprecedented at any point in human history," Altman would later say, "that a technology has improved this much this fast."

At Anthropic, Dario Amodei (OpenAI→Anthropic) and his team had left OpenAI over disagreements about safety, but not about scaling. They believed the capabilities were real. They just wanted to build them more carefully. Constitutional AI, their signature innovation, was itself a scaling solution: train the model to check its own outputs, then scale up the checking.

At Google, the awakening came later but hit harder.

December 2022. ChatGPT had launched weeks earlier and was growing faster than anything in technology history. One million users in five days. A hundred million within two months. Sundar Pichai (Google) declared "code red."

The irony was unmistakable.

"80, 90% of the breakthroughs that underpin the modern AI field today was from Google Brain, Google Research, and DeepMind," Demis Hassabis (DeepMind) would note. The transformer architecture, the "T" in GPT, came from a 2017 paper written entirely by Google researchers. Google had developed BERT. Google had acquired DeepMind. Google had invented the technology.

And yet OpenAI had shipped. Google had not.

The company that created the technology didn't create the product. Research excellence did not equal product execution. Google had everything (the talent, the technology, the compute, the data) and still missed the moment.

Now that had to change.

The Return

Then Sergey Brin (Google) came back.

The Google co-founder had stepped away from day-to-day operations in late 2019. He'd planned to read physics books in cafes, the kind of leisurely retirement a billionaire could afford. That plan lasted about a month.

"There was a guy from OpenAI, this guy named Dan," Brin recalled. "And I ran into him at a little party and he said, 'You know, look, what are you doing? This is like the greatest transformative moment in computer science ever.'"

Brin was a computer scientist at heart, technically still on leave from his Stanford PhD. The rebuke stung. So he started going to the office again, tentatively at first, then with increasing urgency. By the time ChatGPT launched, he was already deep in the code.

"It's been some of the most fun I've had in my life, honestly," he said. "Being a computer scientist, it is the most exciting thing of my life, just technologically."

The pace of change was unlike anything he had experienced, and Brin had experienced the early web.

"I remember using Mosaic and then later Netscape... the what's new page, you know, where it was like, in this last week, these were the new websites. These were the three new sites on the whole internet." He laughed at the memory. "So obviously the web developed very rapidly from there... but the developments in AI are just astonishing by comparison. The web spread, but didn't technically change so much from month to month, year to year. But these AI systems actually changed quite a lot. If you went away somewhere for a month and you came back, you'd be like, whoa, what happened?"

Brin started submitting code, basic stuff at first, access permissions and minor changes. "The code I submitted wasn't very exciting," he admitted. "I think I needed to add myself to the access list." But it sent a message through the organization. Daddy was home. The founder was in the trenches.

There was friction. Google had grown into a bureaucracy that Brin didn't hire, with rules that seemed absurd in retrospect. "I had a big tiff inside the company," he admitted, "because we had this list of what you're allowed to use to code and what you're not allowed to use to code, and Gemini was on the no list."

He pushed back. He went to Sundar. Eventually, it got fixed. But the episode illustrated why Google had hesitated in the first place: a company that couldn't use its own AI for internal coding was a company with institutional caution baked into its DNA.

Now that caution had to be overcome.

The Convergence

One insight gave the believers confidence: everything was converging.

Brin had watched the field evolve for years. "You used to have all kinds of different kinds of models," he observed. "Convolutional networks for vision things. RNNs for text and speech. And all of this has shifted to transformers basically. And increasingly it's also just becoming one model."

The specialized approaches still had their place. AlphaFold for protein structure. Custom architectures for chip design. But the trend pointed toward unification.

"We are generally able to, after we do that, take those learnings and basically put that capability into a general model. So there's not that much benefit. You can get away with a somewhat smaller specialized model, a little bit faster, a little bit cheaper, but the trends have not gone that way."

The implication was significant. If the future belonged to general models rather than specialized ones, then scale and research excellence mattered more than narrow optimization. The company that could train the biggest model would win.

This was the faith. One architecture. One direction. Scale as the universal solution.

The Race

By 2023, the scaling era was in full sprint.

The numbers told the story. GPT-3 had 175 billion parameters. GPT-4 was estimated to be far larger. OpenAI stopped disclosing exact figures, but industry estimates suggested a mixture of experts totaling over a trillion parameters. Each generation cost more to train. Each generation performed better.

"Up until 2020, from 2012 to 2020, it was the age of research," Sutskever reflected. "Now from 2020 to 2025, it was the age of scaling. Because people say this is amazing, you've got to scale more, keep scaling, the one word scaling."

The investment followed. Microsoft invested over $13 billion in OpenAI. Amazon invested $4 billion in Anthropic. Google merged Brain and DeepMind into a single organization with over a thousand researchers. Meta released Llama to commoditize the model layer and ensure everyone could scale.

The reasoning was identical across companies, even as their strategies diverged.

OpenAI bet on closed models and consumer products. Anthropic bet on safety and enterprise. Google bet on integration with its existing products. Meta bet on open source and the social graph.

Different strategies. Same underlying conviction. Bigger was better.

The loss curves didn't bend. The capabilities kept emerging. The scaling laws held.

"You know you'll get something from pre-training."

The Evidence

The evidence seemed overwhelming.

GPT-4 could pass the bar exam. It could score in the 90th percentile on the SAT. It could write code, debug programs, explain complex concepts, generate creative content. Each capability had emerged from scale, not from clever engineering of that specific skill, but from the brute force of training on more data with more parameters.

"This is a model for the first time where I feel like I can ask kind of any hard scientific or technical question and get a pretty good answer," Altman said of GPT-5. He recalled building a simple game called Snake on his TI-83 calculator as a kid, hours of painful programming. "And of course it did that perfectly in like seven seconds."

The experiences accumulated. Developers who had struggled for weeks with particular coding problems watched models solve them in seconds. Researchers who had spent careers mastering narrow domains found chatbots that could engage substantively across all of them.

And it kept getting better. Each new model release brought new capabilities that had not been explicitly trained. Emergence, the researchers called it: abilities that appeared suddenly at certain scales, unpredictable but consistent.

The bitter lesson was proving itself out. General methods that leveraged computation were winning. Human expertise was being superseded.

The Psychology of Belief

What made the scaling faith so powerful was not just the evidence. It was the psychology.

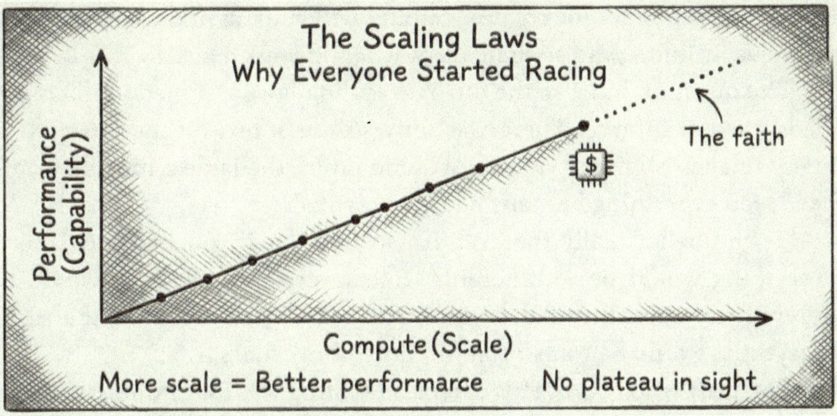

For researchers, scaling offered clarity in a field that had spent decades wandering. The AI winters, periods where funding dried up and progress stalled, had been caused by overpromising and underdelivering. Now there was a clear path forward. Train bigger models. The results would follow.

For companies, scaling offered predictability in a domain that had always resisted it. You could plan a roadmap. You could budget for capability improvements. You could tell investors when the next model would ship and have confidence you could deliver.

For investors, scaling offered a thesis. AI was not magic or luck or dependent on genius researchers who might leave. It was infrastructure. It was capital expenditure. It was the kind of problem that money could solve.

"Companies love this because it gives you a very low-risk way of investing your resources."

The love was mutual. Venture capitalists who had been burned by AI hype in previous decades found themselves pouring billions into companies whose pitch was simple: we will scale faster than anyone else. The scaling laws were the business plan.

The First Doubt

But Sutskever, even as he articulated the faith, was starting to see its limits.

"At some point, though, pre-training will run out of data. The data is very clearly finite. And so then, okay, what do you do next?"

The question hung in the air. The scaling laws assumed infinite fuel. Data was not infinite. There was only so much text on the internet, so many images, so many videos. At some point, the largest models would have seen everything humans had ever created.

"Is the belief really that, oh, it's so big, but if you had 100x more, everything would be so different?" Sutskever asked. "Like, it would be different for sure. But like, is the belief that if you just 100x the scale, everything would be transformed? I don't think that's true."

The man who had bet his career on scaling was questioning whether scaling alone would be enough.

"So it's back to the age of research again," he concluded. "Just with big computers."

But in 2023, few were listening. The race was on. The billions were flowing. The models were shipping.

The era of conviction had reached its peak.

The Church

The metaphor was irresistible.

The AI labs had become cathedrals of compute. Their data centers hummed with thousands of GPUs, each one doing the work that had once required entire university departments. The pilgrims came from Stanford and MIT, from Google and Facebook, drawn by the faith that something unprecedented was being built.

They had their sacred texts: Sutton's essay, the scaling laws paper, Attention Is All You Need. They had their prophets: Hinton, Sutskever, Altman, Hassabis. They had their rituals: the benchmark competitions, the paper releases, the model launches that drew breathless coverage.

And they had their schismatics. Researchers who questioned whether language models could ever achieve true understanding. Safety advocates who warned about the dangers of racing ahead. Critics who saw the scaling faith as a collective delusion, a bubble waiting to pop.

But the schismatics were losing. The models kept getting better. The benchmarks kept falling. The money kept flowing.

"It's crazy," Sutskever mused, "that all of this is real. Like, all this AI stuff and all this Bay Area, yeah, that it's happened, like... Isn't it straight out of science fiction?"

It was. And for a brief moment, it seemed like the science fiction was becoming fact. The scaling laws held. The emergence continued. The capabilities improved.

The faith was being rewarded.

What no one knew, not yet, was that the era of pure conviction was about to end. The believers were about to start fighting among themselves. Not about whether to scale, but about who should control what they were building, and at what cost.

The fracturing was coming.

15

THE FRACTURING

The Schism

The fracturing began inside OpenAI itself.

In December 2020, a group of senior researchers gathered to discuss their growing unease. They had built GPT-2. They had built GPT-3. They had watched language models go from parlor tricks to something that felt genuinely strange: systems that could write, reason, and create in ways that kept surprising their creators.

"There was a group of us within OpenAI," Dario Amodei (OpenAI→Anthropic) later explained, "that in the wake of making GPT-2 and GPT-3, had a kind of very strong focus belief in two things."

The first belief: these systems were going to get much more powerful, much faster than most people expected.

The second belief: the safety work wasn't keeping pace.

Amodei had spent five years as vice president of research at OpenAI. Alongside Ilya Sutskever (OpenAI→SSI), he had helped set the organization's research direction. He remembered the moment his conviction crystallized, when Ilya had explained his philosophy in a single sentence.

"The thing you need to understand about these models," Sutskever told him, "is they just want to learn."

It was a Zen koan, the kind of statement that explained a thousand observations. The models weren't fighting against their training. They weren't resisting. Point them in the right direction, give them enough data and compute, and they simply learned. Get out of their way.

But that power cut both ways. If the models just wanted to learn, they would learn whatever you pointed them at, including things you didn't intend.

"People say we left because we didn't like the deal with Microsoft," Amodei later clarified. "False. Although, there was a lot of discussion, a lot of questions about exactly how we do the deal with Microsoft." The real issue was deeper. "We left because we didn't like commercialization? That's not true. We built GPT-3, which was the model that was commercialized. I was involved in commercialization."

The disagreement was about something harder to articulate. Not whether to build. Not whether to deploy. But *how*, and with what safeguards.

"How can safety not just be something we say because it helps with recruiting?" That was the question Amodei kept returning to. "How do we have a real vision for how to get it right?"

Over many years at OpenAI, he had watched safety become part of the organization's messaging without becoming part of its core identity. The words were there. The commitment felt uncertain.

"It is incredibly unproductive to try and argue with someone else's vision," Amodei concluded. "Take some people you trust and go off together and make your vision happen."

In early 2021, Dario Amodei, his sister Daniela (Anthropic), and five other senior OpenAI employees departed to found Anthropic. They took key researchers from the GPT-3 team. They took their conviction that safety and capability had to advance together, that you couldn't just build powerful systems and figure out alignment later.

But Amodei also knew what worried him most.

"I am optimistic about meaning," he said. "I worry about economics and the concentration of power. That's actually what I worry about more, the abuse of power. And AI increases the amount of power in the world, and if you concentrate that power and abuse that power, it can do immeasurable damage."

The split exposed a paradox at the heart of AI safety.

You can't study what you don't build. If you want to understand how powerful AI systems fail, you need powerful AI systems to study. If you want to develop alignment techniques, you need capable systems to align. Safety research requires capability research.

But building powerful systems creates the dangers you're trying to prevent. Every advance in capability is another step toward systems that might be too capable to control.

Anthropic's answer was what they called "responsible scaling": advance capability and safety in tandem, never deploying a system until you understand its risks, but never stopping capability work because that's the only way to understand what's coming.

Their technical innovation was Constitutional AI, training models to evaluate their own outputs against explicit principles rather than relying solely on human feedback. The idea was to make alignment scalable. You couldn't have humans review every response. But you could train the model to internalize human values and police itself.

Critics saw a contradiction. Anthropic was building frontier AI systems while warning about the dangers of frontier AI systems. They were racing while preaching caution.

But Amodei saw Anthropic as something else entirely: a clean experiment.

"Go off and do a clean experiment and say, 'This is our vision. This is how we're going to do things. Your choice is you can ignore us, you can reject what we're doing, or you can start to become more like us.'"

A race to the top instead of a race to the bottom. Build the systems carefully. Demonstrate that safety and capability aren't mutually exclusive. Pull the industry toward better practices through competition rather than argument.

"At the end," Amodei said, "it's not about one company winning or another company winning. The thing that matters is the ecosystem that we all operate in and how to make that ecosystem better."

By 2026, the clean experiment was showing results. "Anthropic is known to be so culturally deeply committed and organized," Lambert reflected. "We hear so little from them, and everybody at Anthropic seems very aligned."

The contrast with OpenAI's public drama was stark. Where OpenAI had board coups and executive departures, Anthropic maintained operational discipline. Where OpenAI's culture leaked constantly to the press, Anthropic kept its internal workings private. The bet was that culture would compound: that an organization aligned on values could move faster, with less friction, than one perpetually managing internal conflict.

"Anthropic is known for betting very hard on code," Lambert noted. "The Claude Code thing is working out for them."

The AI race wasn't just about models. It was about organizations. And organizations, ultimately, were about culture.

The AI field had fractured. The people who had built the most powerful systems were now split over how, and whether, to continue.

The Speciesist Revisited

The story of OpenAI's founding – Musk's late-night arguments with Larry Page, the "speciesist" accusation, the billion-dollar commitment to keep AI open – was told earlier in this book. But to understand the fracturing, you need to see what happened next.

The exchange that launched it all, according to Musk, deserves one detail we left out. When he pushed back on what he perceived as Page's indifference to human welfare, Musk recounted, the conversation took a turn.

"And I'm like, 'Well, what team are you on, Larry?'" Musk told Lex Fridman. "He's like, 'I'm on team robot.' And I'm like, 'Okay.'"

Team robot. If Musk's account is accurate, it was meant as philosophy. He heard it as warning.

OpenAI was his answer: a nonprofit counterweight to Google's concentrated power. Open source. Transparent. Safety-first. Musk provided over $44 million of the initial funding, recruited Ilya Sutskever from Google in a battle with Demis Hassabis, and helped assemble a team that believed AI development should happen in the open.

For a while, it worked. Papers were published. Code was released. The organization functioned as a research lab with a conscience.

Then the nonprofit model couldn't scale.

Training frontier AI systems cost tens of millions of dollars, then hundreds of millions, then billions. The scaling hypothesis demanded resources that charitable donations couldn't provide.

In 2018, Musk left the board. Reports suggested he had proposed taking OpenAI under his own control, frustrated with what he saw as its inability to keep pace with Google. The rest of the board and staff refused.

Musk departed and reneged on a pledge to deliver $1 billion in additional funding.

The transformation accelerated. OpenAI created a "capped-profit" subsidiary. Microsoft invested $1 billion, then another $10 billion. The organization that was founded to prevent AI from falling into the hands of a single corporation became dependent on the world's largest software company.

By 2023, GPT-4's architecture remained entirely secret.

The "open" in OpenAI had become a historical artifact.

Five Days in November

On November 17, 2023, at approximately noon Pacific time, Sam Altman (OpenAI) received a text asking him to join a Google Meet.

He was watching the Las Vegas Grand Prix. The text came from Ilya Sutskever (OpenAI→Safe Superintelligence), his co-founder, his chief scientist, the man he'd helped recruit from Google eight years earlier. The request wasn't unusual. OpenAI executives talked all the time.

Altman joined the call. The entire board was waiting, everyone except Greg Brockman, the company's president and another co-founder. Sutskever delivered the news: Altman was being fired, effective immediately. The announcement would go public shortly.

Five to ten minutes later, it did.

"The board no longer has confidence in his ability to continue leading OpenAI."

The CEO of the most important AI company in the world, the man who had become the face of the generative AI revolution, who had testified before Congress, who had built ChatGPT into the fastest-adopted consumer product in history, was gone. No warning. No negotiation. Just a Google Meet and a press release.

Within the hour, Greg Brockman learned he was being removed from the board as well. He resigned that evening in protest.

Microsoft, OpenAI's partner, its $13 billion investor, the company that had built its AI strategy around this relationship, learned about the firing "a minute" before the world did.

The AI industry was in chaos. And nobody outside the boardroom understood why.

The specific complaints would emerge only later. In May 2024, former board member Helen Toner broke her silence on The TED AI Show. According to Toner, Altman had not informed the board about ChatGPT's launch before it happened — "we learned about ChatGPT on Twitter." He hadn't disclosed that he personally owned the OpenAI Startup Fund. And in October 2023, a month before the firing, two executives had approached the board with documented concerns — screenshots and records of what they described as problematic interactions and mistruths. "They used the phrase 'alleged psychological abuse,'"

Toner said, "telling us they didn't think he was the right person to lead the company to AGI." OpenAI has disputed elements of Toner's account.

Sutskever had grown increasingly uneasy about OpenAI's direction: the pace of deployment, the commercial focus, the drift from safety research. He had helped establish the superalignment team precisely because he worried that capability was outpacing control. Whatever his reasoning, he supported the board's decision to act.

What followed was chaos.

Friday, November 17: Altman is fired. Brockman resigns. The world learns what happened from a press release and social media posts.

Saturday, November 18: Investors pressure the board to reverse course. OpenAI employees show support for Altman by posting heart emojis on his tweets.

Sunday, November 19: Altman returns to OpenAI headquarters wearing a guest badge. He sets a deadline: reinstate him with a new board, or he walks. The deadline passes. The board names Emmett Shear, former Twitch CEO, as interim leader.

Monday, November 20: Microsoft's Satya Nadella announces that Altman and Brockman will lead a new AI research unit at Microsoft. Nearly all of OpenAI's 770 employees sign a letter threatening to quit and join them unless the board resigns and reinstates Altman.

Among the signatories: Ilya Sutskever himself.

"I deeply regret my participation in the board's actions," Sutskever posted on X.

Tuesday, November 21: OpenAI announces that Altman will return as CEO with a new board. The board members who voted for his removal are out.

Five days. The most dramatic corporate crisis in AI history. And Altman emerged stronger than before.

What the Board Got Wrong

The board had underestimated something fundamental: in the AI race, talent is everything.

OpenAI's employees could walk across the street to any of a dozen well-funded competitors. Google would hire them. Anthropic would hire them. Microsoft had just offered to hire all of them. The researchers who made OpenAI valuable weren't captive; they were the most sought-after people in technology.

When 95% of the company signed a letter threatening to leave, the board's position collapsed. They could fire the CEO, but they couldn't fire the company.

There was also the question of power. Microsoft had invested $13 billion. Whatever the nonprofit charter said about control, the practical reality was that OpenAI's commercial success depended on that relationship. When Nadella offered to absorb the entire operation, the board faced a choice between governance principles and organizational survival.

Principles lost.

The new board, led by former Salesforce CEO Bret Taylor, with former Treasury Secretary Larry Summers, lacked the technical expertise or AI safety background of its predecessor. They were, as Musk's lawyers would later put it, "more experience in profit-centric enterprises or politics than in AI ethics and governance."

They were also, reportedly, "big fans of Altman."

Sutskever stayed at OpenAI after the crisis, but only briefly. His apology had been public. His position had been compromised. The superalignment team he co-led was quietly defunded. By May 2024, he was gone, founding Safe Superintelligence Inc. with a mission to build what OpenAI's name had once promised.

The founding vision had scattered to separate companies, separate missions, separate visions of the future.

The Godfather's Warning

While OpenAI tore itself apart, another fracture was forming, this one between the creators and their creation.

On May 1, 2023, Geoffrey Hinton resigned from Google.

After a decade at the company that had bought his neural network ideas for $44 million, after half a century pushing an approach that most researchers dismissed as a dead end, after vindication so complete it came with a Turing Award and would soon include a Nobel Prize, the godfather of deep learning walked away.

He had one reason.

"I left so that I could talk about the dangers of AI without considering how this impacts Google."

The precise timing was important. There was a conference at MIT. Hinton wanted to speak freely. He couldn't do that as a Google employee.

"You kind of censor yourself," he admitted. "If you work for a big company, you don't feel right saying things that will damage the big company. Even if you could get away with it, it just feels wrong to me."

So the man who had kept faith for fifty years, through two AI winters, through mockery from the establishment, through decades when neural networks were considered a fringe idea, chose to spend his final act warning the world about what he had built.

His transformation had begun the year before, when he realized something about digital intelligence that shook his certainty.

"The closest I had to a Eureka moment was when a Google system called PaLM was able to say why a joke was funny," he recalled. "I'd always thought of that as a kind of landmark. If it can say why a joke's funny, it really does understand."

Then came the second realization, more technical but more troubling.

Biological intelligence is mortal. When you die, your knowledge dies with you. What survives is what you managed to compress into language and teach to others, a lossy, low-bandwidth transmission at best. Humans learn one mind at a time, each generation starting nearly from scratch.

Digital intelligence is different. You can run 10,000 copies of the same AI on 10,000 computers. They can look at different data simultaneously. At the end of the hour, they can share what they've learned, not through the bottleneck of language, but by averaging their parameters directly. The knowledge propagates instantly, perfectly, to all copies.

An AI can learn more in an hour than humanity learns in a century.

"That coupled with realizing why digital is so much better than analog for sharing information," Hinton said. "Suddenly made me very interested in AI safety. And that these things were going to get a lot smarter than us."

Now he was trying to convey this to anyone who would listen. Humanity had never ceded the top of the intelligence hierarchy. We had no experience being second.

His friend Yann LeCun (Meta) thought he was wrong. "We build these things. We're always going to be in control. We'll build them to be obedient."

Others, like Eliezer Yudkowsky – founder of the Machine Intelligence Research Institute and one of the earliest voices warning about AI existential risk – thought Hinton wasn't worried enough. "These things are going to wipe us out for sure."

Hinton tried to stake out a middle position.

"I think both of those positions are extreme. It's very hard to estimate the probabilities in between." He paused. "If I had to bet, I'd say the probability is in between. And I don't know where to estimate it in between."

But he knew one thing: "There's no way we're going to prevent it getting rid of us if it wants to. We're not used to thinking about things smarter than us."

The man who had proven neural networks worked was now asking whether proving them right had been wise.

The Impossible Tradeoffs

There was a deeper problem than any single company's decisions.

"Can you slow it down?" Hinton was asked.

"I don't believe we're going to slow it down."

The reason was structural. Competition between companies within countries. Competition between countries themselves. Even if one organization paused, others would continue. Even if America paused, China wouldn't.

"And if the US slowed it down, China wouldn't slow it down."

In March 2023, the Future of Life Institute had published an open letter titled "Pause Giant AI Experiments." It called for a six-month moratorium on training systems more powerful than GPT-4. Over 30,000 people signed, including many AI researchers.

None of the major AI labs stopped. If anything, investment accelerated. The global AI infrastructure buildout, data centers, chip production, training runs, continued at unprecedented scale.

"They're not going to stop it because it's too good for too many things."

The benefits were real. AI could revolutionize healthcare, education, scientific discovery. It could accelerate drug development, personalize learning, solve problems that had resisted human intelligence for decades. No government was going to halt that progress over speculative risks.

The strange thing about the safety debate was that the people most worried were the people who understood the technology best. Not the critics who thought AI was overhyped. Not the skeptics who doubted current systems were really "intelligent" in any meaningful sense. But the builders themselves.

Hinton, who had pushed neural networks for fifty years, now warning they might surpass and supplant us.

Sutskever, who had bet his career on scaling, now founding a company with "Safe" in its name.

Amodei, who had helped build GPT-3, now leading an organization explicitly focused on alignment.

Bengio (Mila), who had nearly a million citations to his name, now speaking about "the future of my children" as his reason for sounding the alarm.

"The people closest to the creation are the most worried."

This was the pattern with transformative technologies. The nuclear physicists who built the bomb formed the Bulletin of the Atomic Scientists. The geneticists who developed CRISPR called for moratoriums. Now the AI researchers who made it work were warning about what they'd made.

But the warnings weren't stopping anyone. The experiment was underway.

The Open Heretic

The third fracture ran along a different fault line: open versus closed.

In February 2023, Meta released LLaMA. The company positioned it as a research release, available to academics and vetted researchers, not the general public. The weights would be shared carefully, responsibly.

Within days, the weights leaked. Someone posted them to 4chan. Within hours, they were everywhere.

And then something unexpected happened.

Within weeks, an entire ecosystem emerged. Researchers fine-tuned the models for specific tasks. Developers created quantized versions that ran on laptops instead of server farms. Stanford released Alpaca, showing that a relatively small model could match the capabilities of systems costing vastly more to train.

The frontier labs, OpenAI, Google, Anthropic, had built moats around their models. Proprietary weights, API access only, subscription fees. Their technology was powerful but locked away.

Meta had torn the walls down.

Yann LeCun, Meta's chief AI scientist, had been arguing for this approach for years. The Turing Award winner believed that concentration of power, whether in a few companies or a few governments, was the greatest threat the field faced. Not superintelligence. Not rogue AIs. The danger was that a handful of organizations would control humanity's access to thinking machines.

"I see the danger of this concentration of power through proprietary AI systems as a much bigger danger than everything else," he argued. "What works against this is people who think that for reasons of security, we should keep AI systems under lock and key, because it's too dangerous to put it in the hands of everybody."

This was heresy in some circles. Many safety researchers argued that precisely because AI was powerful, access should be restricted.

Releasing model weights to the world meant anyone could use them, including bad actors.

LeCun didn't buy it.

"That would lead to a very bad future in which all of our information diet is controlled by a small number of companies through proprietary systems."

Mark Zuckerberg (Meta) articulated a different framing. "It's only God if only kind of like one company or government controls it," he explained. "If you were the only person who had access to a computer and the internet you would have this inhuman power that everyone else didn't have. But when everyone has it, it makes us all better, but it's also an even playing field."

Meta's embrace of open source wasn't just philosophy. It was strategy. Zuckerberg didn't need to sell AI subscriptions. Meta made its money from advertising, from understanding users well enough to show them relevant ads. If AI became a commodity, if everyone had access to powerful models, Meta would be fine. Their advantage wasn't the model itself. It was the social graph, the user data, the platform that three billion people used every day.

The calculation was elegant: give away the model, keep the social graph.

This wasn't charity. It was competition by other means.

But would it last?

By early 2026, researchers were asking whether Meta would continue the Llama series. The models had become critical infrastructure for the open-source AI community. Thousands of projects, hundreds of startups, entire research agendas depended on competitive open weights from a company with the resources to produce them.

"Llama was the most focused expression of the organization," Lambert explained. "I don't see Llama being supported to that extent."

When asked directly whether there would be a Llama 5, Lambert—who does not speak for Meta—was blunt: "Not an open-weight one."

If true, the implications were significant. The open-source community had built on the assumption that competitive open models would continue. A retreat by Meta would leave Chinese labs as the primary source of frontier open weights.

The irony was sharp. America had pioneered open-source AI. China might inherit it.

The European Alternative

In Paris, another heretic was making a different bet.

Arthur Mensch (DeepMind→Mistral) had been at DeepMind, watching the field transform from the inside. By 2023, he had grown frustrated with the direction it was taking.

"Around 2020, some companies started to be quite ahead in some field and realized that some value could be accrued," he observed. "And then at that point, opacity made it back to the field." He found this deeply troubling. "We think it's really damaging for the science to actually move into such an opaque regime, where you have a few companies basically doing the same thing, just not communicating about it, spending billions of compute doing exactly the same thing."

Mistral, the company he co-founded, was built on a different premise. Open source wasn't just better for democracy; it was better for AI progress itself.

"We think it's too early, and we think it's really damaging for the science. The technology we're looking at is not working completely yet. It doesn't reason well. Memory mechanisms are not very well understood. There's a lot of biases. It's incomplete. There's many things to be done."

His complaint had bite because Mensch wasn't arguing from the outside. He'd seen how the frontier labs worked from within.

"We still need to invent new techniques. And how are we going to invent new techniques if nobody is speaking about it, when in order to invent new techniques, you need to still spend some large amount of money to actually try things at scale?"

Mistral released its 7B parameter model in September 2023, tiny by frontier standards but punching far above its weight class. The model ran on laptops. It cost almost nothing to serve. And it outperformed models many times its size.

But Mensch was also pragmatic about safety arguments that seemed designed to protect incumbents.

"The current discourse – the really big tech companies are claiming that open source AI is dangerous, and it feels like really a form of regulatory capture. They want regulators to step in so that they can constrain innovation and kind of control an industry."

The European dimension mattered. "France, UK, Poland are very good at training mathematicians. And as it turns out, mathematicians are very good at making AI." Mensch saw an opportunity to build something important outside Silicon Valley's gravitational pull, and to prove that frontier AI didn't require American scale.

The Fracture Lines

By the end of 2023, the scaling believers had splintered along three fault lines.

Safety versus speed. Anthropic and Safe Superintelligence said: slow down, build carefully, solve alignment before capability outpaces control. OpenAI and xAI said: build fast, iterate, learn from deployment. Google tried to do both and struggled to do either.

Open versus closed. Meta and Mistral said: democratize access, prevent concentration, let a thousand experiments bloom. OpenAI and Anthropic said: these systems are too powerful to release freely, control access, prevent misuse.

American versus everyone else. Silicon Valley had dominated the first decade of deep learning. Now Europe was building alternatives. China was racing ahead. The assumption that AI was an American monopoly was starting to crack.

But here was the strange thing: on the fundamental question, the question of whether scaling worked, whether bigger was better, whether the path to intelligence ran through more data and more compute, everyone still agreed.

Anthropic built larger models while preaching safety. Meta released open weights while racing to frontier scale. Mistral celebrated efficiency while planning billion-dollar training runs. Even Hinton, warning from retirement, didn't question that the technology worked. He questioned whether working was wise.

The fractures were real. The debates were fierce. Friendships ended. Companies split. Lawsuits were filed.

But all the combatants remained believers in the same underlying faith. The scaling laws held. The capabilities kept emerging. The loss curves didn't bend.

The fight wasn't about whether to build the cathedral. It was about who should control it, who should have access, who should bear the risks.

And while they fought among themselves, a lab in China was about to prove them all wrong.

16

THE EASTERN FRONT

The Shock

In late 2024, a Chinese AI lab began making waves that would soon become a tsunami.

DeepSeek-V3, released in December, had already turned heads. A model trained for under $6 million that competed with systems costing hundreds of millions. But it was DeepSeek-R1, released in January 2025, that truly shook the industry. The reasoning model matched OpenAI's o1 on key benchmarks. Its claimed training cost, according to the company's technical report: $5.6 million. Independent analyses suggested the true figure was even lower, perhaps $294,000 for the final training run alone.

OpenAI had spent hundreds of millions. Meta had spent billions on infrastructure. The entire premise of frontier AI development was that capability required capital at a scale only the largest companies could afford.

DeepSeek had just proved that premise wrong.

Within days, Nvidia's stock dropped 17%, erasing nearly $600 billion in market value. It was the largest single-day decline for any company in U.S. stock market history. The DeepSeek app surged to number one in Apple's U.S. App Store, ahead of ChatGPT. Marc Andreessen called it

"AI's Sputnik moment."

The export controls were supposed to cripple Chinese AI development. The United States had banned the sale of Nvidia's most powerful chips, the H100s and A100s, to Chinese companies in October 2022. The logic was straightforward: limit compute, limit capability. Without the hardware, China couldn't compete.

Instead, they had found a different way.

The Quant Trader

Liang Wenfeng had never worked in artificial intelligence.

Born in Guangdong in 1985, he studied electronic and information engineering at Zhejiang University. After graduating, he went into quantitative trading, building algorithms that could find patterns in financial markets. In 2016, he co-founded High-Flyer, a hedge fund that used AI to trade. It became one of China's largest quant funds, managing over $8 billion at its peak.

But Liang was obsessed with something bigger than markets. He saw what was happening with large language models: the emergence of systems that could reason, that could understand, that might someday think. And he believed that China needed to participate in that revolution, not just follow it.

"For years, Chinese companies have been accustomed to leveraging technological innovations developed elsewhere and monetizing them through applications," he said in a July 2024 interview. "But this isn't sustainable. This time, our goal isn't quick profits but advancing the technological frontier."

In 2023, he founded DeepSeek with registered capital of just $1.4 million. The team was small, fewer than 200 researchers, mostly fresh graduates from top Chinese universities. They had no overseas experience, no pedigree from Google or Meta or OpenAI.

What they had was chips. Before the export controls took effect, High-Flyer had stockpiled 10,000 Nvidia A100s. It was the only private company in China with that scale of compute outside the tech giants. Liang hadn't bought them for business reasons. He'd bought them out of curiosity: a quant trader's instinct that AI was going to matter.

Now that curiosity would pay off.

The Study Group

To understand what DeepSeek achieved, you have to understand how differently China approached the AI race.

Kai-Fu Lee had spent forty years watching both ecosystems – AI at Apple, Google, Microsoft Research China. When the generative AI revolution exploded, he was one of the few people who understood both sides of the Pacific well enough to see what was really happening.

"What drives American generative AI companies," Lee observed, "is that each of the four big ones, OpenAI, Anthropic, Google, and xAI, wants to be the one that cracks AGI. And they think with AGI comes enormous power and the ability to not only win but effectively cause everyone else to lose in the business sense."

But the Chinese companies saw the game differently.

"The American model is like a genius who wants to win the Nobel Prize and four geniuses competing to win," Lee explained. "The Chinese companies are more of a study group. They compete with each other, but they know they're learning from each other. They're willing to let each other learn from each other because they figure everybody will get there. And the winner will be the one who develops a business model that can make money."

This was a fundamentally different bet. The Chinese labs believed that no matter who crossed the finish line first, everyone else would follow within six to twelve months. If that was true, the race wasn't about achieving AGI first. It was about being efficient enough to profit from it when everyone had it.

"If everybody co-arrives," Lee asked, "what's the point of Stargate? What's the point of spending a trillion dollars to do experiments to learn? Let somebody else spend the money and learn from them."

Efficiency as Virtue

DeepSeek's breakthrough wasn't about having more resources. It was about using fewer resources better.

The innovations were technical: mixture-of-experts architectures that activated only a fraction of the model's parameters at any given time.

Novel optimization techniques. Clever approaches to reinforcement learning that reduced the need for expensive human feedback.

But the real innovation was cultural. DeepSeek treated efficiency not as a constraint but as a design principle.

"More investment does not necessarily lead to more innovation," Liang observed. "Otherwise, large companies would take over all innovation."

This was heresy in Silicon Valley, where the assumption was that capability scaled with compute, and compute scaled with capital.

DeepSeek suggested a different equation. What if capability scaled with ingenuity? What if the right algorithm could substitute for the wrong hardware?

The R1 Earthquake

DeepSeek-V3, released in December 2024, demonstrated what efficiency gains looked like in practice. The model was trained on 2,048 Nvidia H800s, chips designed specifically for the Chinese market after the export controls, less powerful than the banned H100s. The publicly disclosed training cost, by DeepSeek's accounting: $5.576 million.

Meta's comparable model, Llama 3.1, had used roughly ten times the computing power. OpenAI's models were estimated to cost hundreds of millions. And yet V3 performed competitively on benchmarks. Not quite as good as GPT-4o on everything, but close enough to matter.

Then came R1.

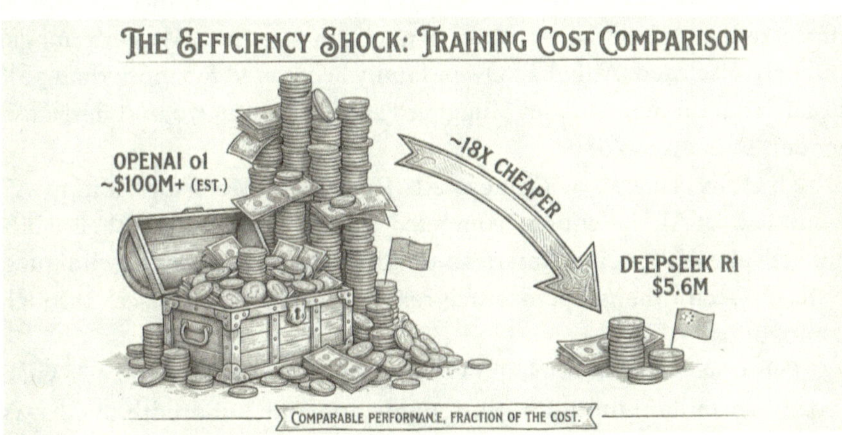

DeepSeek-R1 was a reasoning model, designed to think step by step before answering. It articulated its reasoning process, showing its work the way a student might on a math test. And on certain benchmarks (mathematical reasoning, coding tasks) it matched or exceeded OpenAI's o1.

The technical paper explained how. Group Relative Policy Optimization, a reinforcement learning technique that let the model learn from comparing its own outputs rather than relying on expensive human feedback. Distillation from larger models. Clever training recipes that squeezed more capability from less compute.

But the paper also raised questions. Some analysts suggested DeepSeek had access to more chips than it claimed, perhaps 50,000 H100s obtained through gray markets despite the export controls. Others suggested they had "distilled" knowledge from OpenAI's models, using ChatGPT's outputs to train their own systems.

DeepSeek denied nothing and explained little. The opacity was itself a message: China didn't need to play by American rules to compete.

The Ecosystem Emerges

Step back, and the pattern was unmistakable.

China hadn't produced one competitor to American AI. It had produced an ecosystem.

Moonshot AI released Kimi K2, a model with one trillion parameters, training cost approximately $4.6 million. Within 48 hours, it became the most downloaded model on Hugging Face. ByteDance's Doubao processed 4 trillion tokens per day, more than any Western model publicly disclosed. Alibaba's Qwen family accounted for more than 30% of all model downloads on Hugging Face, with over 170,000 derivative models built on top of it.

The "Six Tigers" of Chinese AI (DeepSeek, Moonshot, Zhipu AI, MiniMax, 01.AI, Baichuan) competed fiercely with each other while collectively challenging American dominance. They shared techniques, built on each other's open-source releases, drove down prices through competition.

The price war was vicious. When DeepSeek released V2 at 1 yuan ($0.14) per million input tokens, roughly one four-hundredth of GPT-4's cost at the time, competitors scrambled to match. Within weeks, ByteDance, Alibaba, and others had slashed prices. The cost of AI inference collapsed.

"We never intended to be a disruptor; it just happened by accident," Liang said of the price war. "We were simply following our own pace, calculating costs, and setting prices accordingly."

But the disruption was real. What had seemed like a sustainable moat for American companies, the ability to charge premium prices for premium capability, was eroding from both ends. Chinese models were getting better. Chinese prices were getting lower. The squeeze was on.

The Open Source Gambit

But why release the models at all?

The American assumption was that open weights were a competitive liability. If you gave away your technology, competitors could copy it. Closed models, with their API revenue and proprietary advantages, seemed like the obvious business model.

The Chinese labs saw it differently.

"A lot of US tech companies won't pay for an API subscription to Chinese companies for security concerns," Lambert pointed out. "This has been a long-standing habit in tech."

The insight was strategic. American enterprises would never trust Chinese servers with their data. Government contracts were off-limits. Financial services, healthcare, defense: entire sectors were inaccessible through the traditional API model.

But open weights changed the calculation.

"The people at these companies then see open weight models as an ability to influence and take part of a huge growing AI expenditure market in the US."

If you couldn't sell the service, you could give away the model. Let American companies download it, run it on their own infrastructure, build products on top of it. You wouldn't capture the revenue directly. But you would shape the ecosystem. You would establish dependencies. You would prove that Chinese AI could compete.

By early 2026, the strategy was working. DeepSeek's models were being downloaded millions of times. Alibaba's Qwen had become the foundation for over 170,000 derivative models. The Chinese labs couldn't sell to American enterprises. They were doing something potentially more valuable: they were becoming the infrastructure American developers built on.

The irony was sharp. The United States had implemented export controls to prevent China from accessing American chips. China had responded by giving away the software that ran on whatever chips they had.

The Export Control Paradox

The irony haunted American policymakers. The export controls were supposed to cripple Chinese AI. Instead, they may have forced innovation.

Kai-Fu Lee saw it clearly: "It's kind of sanctioning something that Chinese companies aren't likely to do anyway and probably can't afford to do anyway." The trillion-dollar training run was an American obsession. China had never planned to compete on scale.

DeepSeek had to become efficient because it had no choice. The constraint became a competitive advantage. Japanese automakers had become efficient because of oil shocks. Israeli tech had flourished despite a small domestic market. The United States had just handed China a very significant constraint.

The skeptics – Musk, Scale AI's Alexandr Wang – suspected DeepSeek had more chips than claimed. But even if true, that raised its own uncomfortable question: if the export controls could be circumvented, what exactly had they accomplished?

The New Landscape

By mid-2025, the assumptions that had governed the AI race lay in pieces. More compute wins? DeepSeek proved otherwise. America leads? The gap had narrowed dramatically. Closed beats open? Open models from China and Meta were matching closed ones on many tasks. Concentration is inevitable? The Six Tigers suggested a future of distributed capability rather than monopoly.

None of this meant China would win. American labs remained

formidable. But the comfortable certainty – the assumption that resources guaranteed outcomes – had evaporated.

And a deeper question was emerging.

What if the fuel was running out?

The scaling thesis assumed infinite data. Train on more text, more images, more video, and the models would keep improving. The loss curves would keep falling. The capabilities would keep emerging.

But the internet was finite. There was only so much human-generated text. Only so many images. Only so many videos. At some point, the largest models would have consumed everything humans had ever created.

And then what?

The doubters were gathering.

INTERLUDE: THE GAME THAT LEARNED TO WIN

The story pauses here.

The previous interludes explained how neural networks see (backpropagation) and how they read (transformers). This one explains how they might learn to act.

The distinction is simple: everything we've built so far learned by reading. What comes next learns by playing.

TWO WAYS TO LEARN
THE READER
(Language Model Approach)
Jump at the third platform, then dodge left... Sounds like an expert.
WALKTHROUGHS
STRATEGY GUIDES
GAMEFAQs
FORUM POSTS
BEST BUILD FOR LEVEL 7
COMPLETE GAMEPLAY MANUAL
EVERY GUIDE EVER WRITTEN
WALKTHROUGHS
GAMEFAQS
FORUM POSTS
Predicting words ≠ Understanding consequences
THE PLAYER
(Reinforcement Learning Approach)
GAME OVER
GAME OVER
GAME OVER
GAME OVER
GAME OVER
SCORE: 100 → 500
SCORE: 2000
SCORE: 10000
LIVES:
TRY
FAIL
ADJUST
TRY
Experience
No tutorial
No instructions
Just try and see what happens
Dies a thousand times
What humans WROTE
What actually HAPPENS
Attempt 1
Attempt 47
Attempt 203
Predicts what a player would SAY
THE REWARD SIGNAL
Discovers what actually WORKS
VERDICT: Never played
WIN → +1
LOSE → -1
The reward signal: just a number
VERDICT: Learned from experience
"Mimicking what people say ≠ Figuring out what to do."
The agents of today predict. The agents of tomorrow must experience.

THE PLAYERS AND THE READERS

Open a terminal. Type a few words describing what you want to build. Watch as an AI writes code, runs it, hits an error, debugs itself, tries again. Minutes later, a working program exists that would have taken you hours.

The tools have names: Claude Code, Cursor, Codex, Gemini. The industry calls them "coding agents." One researcher describes the experience differently: "It's a little spirit that lives on your computer."

These agents are impressive. For code, for research, for tasks that can be verified, they work spectacularly.

But ask them to learn from their mistakes across sessions. Ask them to remember what worked last time. Ask them to pursue a goal over days instead of minutes. Ask them to understand what happens when you drop a ball.

The illusion breaks.

Here's why: these agents are *readers*. They learned everything they know from text – billions of pages of human writing, ingested and compressed into statistical patterns. They can describe the world with extraordinary fluency. They have never touched it.

What follows is the story of the *players* – systems that learn not from reading, but from doing. And why a growing number of researchers believe that playing, not reading, is the missing piece.

THE GAME

You're dropped into a video game you've never played.

No tutorial. No instructions. Just a controller and a screen. Something moves. You press a button. Something else happens. You don't know the rules. You don't know the goal. You don't even know what the buttons do.

So you try things.

Press A. A character jumps. Press B. Nothing obvious happens. Move the stick left. You walk off a cliff. Game over.

Try again. This time, don't walk off the cliff. Explore a little longer. Find something that glows. Touch it. A number increases. Press B near an enemy. Something fires. The enemy disappears. The number increases again.

Slowly, painfully, through countless deaths and restarts, you learn. Not from a manual. Not from watching someone else play. From *experience*. From trying things and seeing what happens.

This is how children learn to walk. How animals learn to hunt.

This is *reinforcement learning*.

THE READER'S VERSION

Now imagine a different approach to the same game.

Instead of playing, you read every walkthrough ever written. Every forum post. Every strategy guide. Every transcript of someone describing their gameplay. You ingest the entire internet's worth of text about this game – and about every other game ever made.

Now you can describe how to beat any level with fluency that sounds like expertise. You can explain boss strategies, recommend character builds, debate which ending is best. Ask you about the game, and you'll respond like a veteran player.

But here's the thing: you never actually played.

You predict what a player would *say*, not what would actually *work*. You can quote the walkthrough about jumping at the third platform – but you've never felt the timing, never learned the rhythm through failure, never discovered that the walkthrough was wrong about the shortcut on level seven.

Richard Sutton, the pioneer of reinforcement learning whose full story we'll meet in the next chapter, puts it bluntly: "Large language models are about mimicking people, doing what people say you should do. They're not about figuring out what to do."

Readers predict sentences. Players discover consequences.

THE SCORE

What makes playing possible? Feedback.

In a video game, feedback is everywhere. You collect a coin: the score increases. You touch an enemy: you lose a life. You reach a checkpoint: progress saves. The game is constantly telling you –

through numbers, sounds, visuals – whether you're doing well or poorly.

In reinforcement learning, this feedback is called the *reward signal*. It's just a number. Plus one if you win, minus one if you lose. Points for collecting treasure, penalties for taking damage.

David Silver, who led the AlphaGo project at DeepMind, describes the core idea simply: "You basically give the outcome of your game a number. And we say plus one if you win and minus one if you lose. And then we train the system to basically do more of the things that get more reward."

That's it. That's reinforcement learning at its heart. Try things. Get feedback. Do more of what works.

The concept is ancient. Psychologists call it operant conditioning. Dog trainers call it positive reinforcement. Parents call it consequences. What's new is using this feedback to train neural networks, adjusting millions of parameters, nudge by nudge, until the system learns to maximize its score.

Readers have no score. They predict what a human *would* say next. But there's no right answer – just statistical likelihood. Players have a score. They win or they lose. There's ground truth.

THE DILEMMA

Even with a score, learning by playing isn't simple.

Imagine you've found one strategy that works. Spamming the attack button beats most enemies. You win more often than you lose. Good enough, right?

But maybe there's a better strategy: a combo move, a timing trick, a completely different approach. You'll never find it if you keep doing what you already know works.

This is the *exploration-exploitation tradeoff*: the tension between doing what works (exploitation) and trying things that might work better (exploration).

Explore too little, and you get stuck with decent strategies that prevent you from finding great ones. Explore too much, and you waste time on random experiments that lead nowhere.

Every player faces this dilemma. Every person faces it too. Do you order the dish you know you like, or try something new? Do you refine your current approach, or start over?

The systems that learn best find ways to balance both.

THE PLAYER WHO THREW AWAY THE MANUAL

Games became the proving ground for players – systems that learn by playing. The feedback is perfect: you win or you lose.

In 2013, a startup called DeepMind taught a system to play classic Atari games using nothing but pixels and the score. No hand-coded features. No human expertise. Just raw sensory input and the reward signal. The system played millions of games and learned dozens of them at superhuman levels. Google acquired DeepMind the following month.

But the breakthrough that matters most came in Go.

In March 2016, DeepMind's AlphaGo defeated one of history's greatest Go players, four games to one. Move 37 became legendary – a stone placed where no human would ever play, so unconventional that commentators assumed it was a mistake. It wasn't. The machine had found something humans had missed in thousands of years of play.

AlphaGo started as both a reader and a player. It read a database of professional games – centuries of human expertise – then improved through self-play, using RL to discover strategies beyond human knowledge.

But the next version, AlphaZero, did something shocking. It threw away the manual entirely.

"What we discovered," Silver recalls, "was that the human data wasn't necessary. That you could actually throw out the human moves altogether."

AlphaZero started from nothing. No database. No human expertise. Just the rules and a score: win or lose. It played itself millions of times, and it didn't just match AlphaGo – it surpassed it.

"The resulting program not only was able to recover this level of performance, it actually worked better and was able to learn even faster."

The reader, it turned out, was holding the player back. The accumulated wisdom of centuries was a ceiling, not a floor.

THE WALKTHROUGHS RAN OUT

This result haunts the language model era.

The readers – ChatGPT, Claude, Gemini – trained on the sum of human text. Every book, article, forum post, chat transcript. They became uncannily fluent at producing text that sounds like something a human would write.

But human text is finite. There are only so many walkthroughs. The internet has been scraped. The books have been ingested. Researchers call it the "data wall." Synthetic data can extend the runway, but eventually you're training on your own outputs – reading your own walkthroughs – and that leads to strange loops and degradation.

David Silver's analogy: human data is like fossil fuels – we mined it, burned it in our models, and got a certain level of performance for free. But fossil fuels run out. "I think that's what reinforcement learning is," he says. "It's the sustainable fuel – this experience that it can keep generating and using and learning from."

The walkthroughs ran out. The game is still being played.

THE REAL GAME

But here's the problem that kept RL in the backwater for decades: games are clean. The real world is messy.

In Breakout, the score tells you exactly how well you're doing. In life, the score is unclear. What's the reward signal for writing a good essay? For making a wise career choice? For being a good parent? The feedback is delayed, ambiguous, contradictory.

And there's the sample efficiency problem. Atari systems needed millions of games to learn what a child grasps in minutes. You can play a million games of Go in a computer. You can't crash a million cars to teach one to drive.

Andrej Karpathy, who taught neural networks in the previous interlude, frames what's missing:

"We're not actually building animals. We're building ghosts."

Ghosts: readers who have consumed every walkthrough ever written, who can describe any game with fluency that sounds like mastery, but who have never held a controller. Ghosts who can write poetry about heartbreak without having a heart. Ghosts who can describe how to ride a bicycle without ever having a body.

Animals are different. Animals are players. A zebra can stand minutes after birth. A squirrel learns which branches hold weight by trying them. The feedback is immediate and grounded in reality: you fell or you didn't.

Richard Sutton's thought experiment clarifies what the ghosts still lack:

"If we understood a squirrel, we'd be almost all the way there."

A squirrel has never read a walkthrough. It plays the game of the real world – trying branches, feeling them bend, learning from consequences. That's what the builders of the next era are reaching for: AI that learns not from what humans *said* happened, but from what actually *does* happen.

FROM READERS TO PLAYERS

For years, the reader-builders and the player-builders worked in separate worlds. Different conferences, different heroes, different bets on what would matter.

Then, in 2022, they started to merge.

OpenAI published a technique called RLHF – Reinforcement Learning from Human Feedback. The idea: you can't write a mathematical reward function for "be helpful." But you can ask humans which of two responses they prefer, build a model that predicts those preferences, and train against that model.

The result was ChatGPT. The readers had acquired a score. They weren't just predicting text anymore – they were learning to be helpful, to follow instructions, to refuse harmful requests. In a limited sense, the readers had started to play.

The same year, Anthropic published Constitutional AI, showing you could use one AI's feedback instead of human feedback. The readers and the players were building on each other's work. (The binary is a simplification, of course – most frontier systems now combine both approaches in various proportions. But the distinction clarifies what each contributes.)

But this was still a reader learning from human preferences about text. Not a player learning from the world.

The researchers we meet next believe something more is needed. Not just readers with scores – but genuine players. Systems that interact with reality, learn from consequences, and build models of how the world works.

THE NEXT LEVEL

The researchers we meet next believe the second path is essential. Richard Sutton thinks intelligence requires playing, not reading. Yann LeCun thinks it requires understanding the physical world, not just predicting words. Fei-Fei Li thinks it requires spatial reasoning – seeing and acting in three dimensions. David Silver thinks the era of reading is ending and the era of playing is beginning. John Carmack left a legendary career in games to pursue agency itself.

They are all reaching for the same thing: AI that doesn't just describe the world, but understands it. AI that doesn't just predict what humans would say, but knows what would actually happen.

The agents of today are readers with tools attached. The builders of tomorrow are trying to teach them to play.

ACT IV: THE HERETICS RETURN

THE QUESTIONING (2024–?)

The builders themselves start questioning. The prophets of scaling announce limits. The case that prediction isn't enough. And as the pioneers debate what comes next, the tools they built become available to everyone.

17

THE DOUBTERS

In December 2024, Ilya Sutskever (OpenAI→Safe Superintelligence) stood before the audience at NeurIPS, the annual gathering of the machine learning faithful, and announced that the age he had helped create was ending.

"Pre-training as we know it will unquestionably end."

The conference hall went quiet. Four thousand researchers, engineers, and executives had come to Vancouver to celebrate their field's triumph. ChatGPT had crossed three hundred million users. Nvidia had become the most valuable company on Earth. AI was the most important technology of the century.

And now the man who had done more than almost anyone to make it happen was telling them the fuel was running out.

"The data is not growing," Sutskever continued, "because we have but one internet."

He let the phrase land.

"You could even say that data is the fossil fuel of AI. It was created somehow, and now we use it, but we've achieved peak data. And there'll be no more."

Peak data. The industry had built trillion-dollar valuations on the assumption that scale was infinite, that there would always be more text, more images, more videos to train on. The scaling laws promised that capability grew predictably with data, compute, and parameters. Keep scaling. The results will follow.

But the internet was finite. There was only so much human-generated text. At some point, the largest models would have read everything humans had ever written.

Then what?

The Prophet

To understand why this mattered, you have to understand who was saying it.

Ilya Sutskever had been present at every critical moment of the AI revolution. He had co-authored AlexNet, helped build the GPT series at OpenAI, and believed so completely in scaling that he'd staked his career on it.

He was not a skeptic. He was the opposite of a skeptic. He was the prophet who had converted the industry to the scaling faith.

"Scaling is just one word," Sutskever had once explained, "but it's such a powerful word because it informs people what to do. They say, okay, let's try to scale things."

And they had. Following the roadmap Sutskever had helped draw, labs around the world had poured billions into training larger and larger models. OpenAI scaled to GPT-4. Google scaled to Gemini. Anthropic scaled to Claude. Meta scaled to Llama. The loss curves kept falling. The capabilities kept emerging. The scaling hypothesis was validated at every turn.

Until it wasn't enough.

"At some point, though, pre-training will run out of data," Sutskever admitted. "The data is very clearly finite. And so then, okay, what do you do next?"

He offered possibilities: reinforcement learning, synthetic data, new training recipes. But his framing was stark. The era that had defined his career was giving way to something else.

"Up until 2020, from 2012 to 2020, it was the age of research," he said. "From 2020 to 2025, it was the age of scaling. But now the scale is so big. Is the belief really that if you just 100x the scale, everything would be transformed? I don't think that's true."

The man who had bet his career on scaling was announcing its limits.

"So it's back to the age of research again. Just with big computers."

The Godfather's New Act

Sutskever's NeurIPS speech was not the first sign of doubt among the builders. Eighteen months earlier, Geoffrey Hinton had resigned from Google to speak freely about the dangers he saw.

What drove Hinton wasn't just the abstract possibility of superintelligence. It was a specific insight about the nature of digital intelligence, one that had crystallized in his final months at Google.

The realization began with humor. Google's PaLM system had explained why a joke was funny. Hinton had always considered this a landmark: if a system could unpack humor, with its layers of context and surprise, it understood something genuine about human experience.

But the deeper insight was structural. Hinton had been exploring analog computation as an energy-efficient alternative to digital systems. The research forced him to confront a fundamental asymmetry between biological and digital minds.

Biological intelligence is mortal. When you die, your knowledge dies with you. What survives is what you managed to compress into language and teach to others, a lossy, low-bandwidth transmission at best.

Digital intelligence has no such limits. Ten thousand copies of the same AI can examine different data simultaneously, then merge their learning by averaging parameters directly. No bottleneck. No translation loss. Knowledge propagates instantly, perfectly, to all copies.

"GPT-4 has thousands of times more knowledge in about 2% of the weights," Hinton calculated, comparing a trillion-parameter model to the human brain's hundred trillion connections. The machines were more efficient learners. And they had a superpower humans lacked: perfect knowledge sharing.

"And that's why GPT-4 knows so much more than a human. It wasn't one model that did it. It was a whole bunch of copies of the same model running on different hardware."

This was why the stakes had changed. Digital intelligence would not stay at human level. The architecture guaranteed it would surpass us, probably soon. And we had never been second before.

The Godfathers Divided

The three men who shared the 2018 Turing Award for deep learning had traveled together through AI winters and vindication. By 2024, they had split.

Hinton warned that AI might be the last technology humans ever invent, because what comes after might not need us.

Yann LeCun (Meta→Advanced Machine Intelligence) dismissed the concerns as overblown. "We build these things. We'll build them to be obedient."

Yoshua Bengio (Mila) worried, but for different reasons than Hinton.

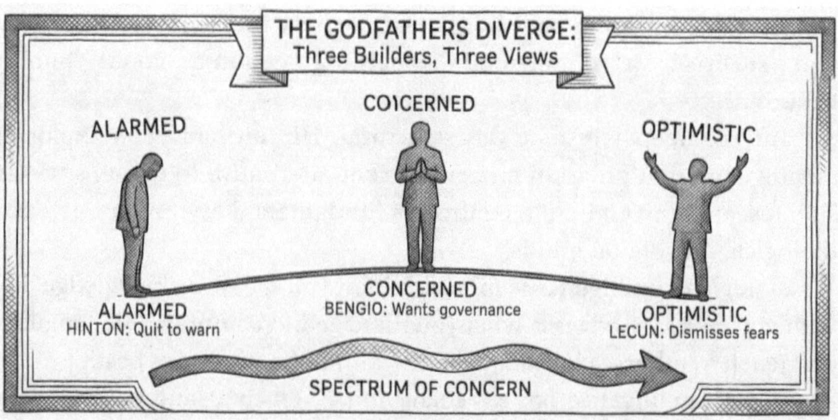

The division ran deep despite decades of collaboration. They had published together, shared the conviction that neural networks would eventually work, and accepted the Turing Award on the same stage. Now they could not agree on what to do with their success.

LeCun's confidence was sincere. He believed the danger was not superintelligence but concentration: a handful of companies controlling how everyone thinks. He advocated for open source precisely because he thought the technology was controllable. "I believe that people are fundamentally good. And so if AI, especially open source AI, can make them smarter, it just empowers the goodness in humans."

Hinton thought this was optimistic to the point of naivete.

"Some people say it's like less than 1 percent," he noted, referring to the probability of existential risk. "Other people, like Yudkowsky, say, no, no, no, these things are going to wipe us out for sure. I think both of those positions are extreme. It's very hard to estimate the probabilities in between."

When pressed on who was right, Hinton refused to commit.

"I simply don't know."

The Third Voice

Yoshua Bengio had taken a different path.

Unlike Hinton, he had not left his position. Unlike LeCun, he had not dismissed the concerns. He had experienced what he called "an emotional pivot."

"Shortly after GPT came out, after a bit of enthusiasm, I realized that we had passed the milestone that Turing, one of the founders of computer science, thought would be basically reaching human-level intelligence and potentially endangering our future."

The realization wasn't purely intellectual. It was personal.

"The future of my children, the emotional reason why I took this so seriously and not just some intellectual question, was such that I couldn't just sit and continue as before."

His colleagues weren't pleased. When Bengio started speaking publicly about risks, he faced pushback from people who felt he was undermining their field, tarnishing the reputation of work they had spent decades building.

"Personally, when I started speaking out, my colleagues were not very happy that I was talking about our field in this way. But I felt that I had no choice."

What worried him most was the trajectory. Benchmarks that had seemed secure for decades were falling in months. Capabilities were emerging faster than anyone could track, let alone understand.

"The biggest mistake that a lot of people make, including AI researchers, but especially policymakers, business people, ordinary citizens, the biggest mistake is to think of AI as we see it now." The data showed capabilities growing relentlessly. "The important point is they're moving fast."

Bengio focused on a specific danger: the push toward AI agents. Systems designed not just to answer questions but to act in the world.

"AI systems that act in the world towards achieving goals, which of course has a lot of economic value because companies wanna replace people by machines, it's going to increase productivity and probably also have effect on society."

Effect on society. A careful understatement for what might be the most significant economic transformation in human history.

"Currently, AI systems are way worse than humans at dealing with tasks that take a lot of time to plan," he acknowledged, "but their ability is growing exponentially fast, and they'll catch up within a few years."

Not decades. Years.

His proposal was radical: don't build agentic AI at all.

"By training an AI to be as good a model of the world as possible, we get something that will not be agentic." Like the laws of physics: understanding without desire, prediction without goal-seeking. "Those laws of physics themselves are not agentic. They don't care about you, they don't care about me, they don't care about our petty goals on this planet."

A scientist AI. An oracle. Something that understood everything but wanted nothing.

Two of the three godfathers had become their own creation's most prominent skeptics. The third continued building.

The Pattern

There was a pattern in who worried most.

Not the critics who thought AI was overhyped. Not the skeptics who doubted current systems were really "intelligent" in any meaningful sense. But the builders themselves.

Hinton, warning from retirement after fifty years of pushing neural networks.

Sutskever, founding a company called Safe Superintelligence Inc., with "Safe" in the name.

Dario Amodei, who had helped build GPT-3, now leading Anthropic with alignment as its explicit focus.

Bengio, with nearly a million citations to his name, speaking about "the future of my children" as his reason for sounding the alarm.

The pattern was consistent. The people closest to the creation were the most worried.

This was not unprecedented. The nuclear physicists who built the bomb formed the Bulletin of the Atomic Scientists. The geneticists who developed CRISPR called for moratoriums. Now the AI researchers who made the current revolution possible were warning about what they'd made.

"I don't feel particularly guilty about developing AI like 40 years ago," Hinton said. "Because at that time we had no idea that this stuff was going to happen this fast. We thought we had plenty of time to worry about things like that."

The time had arrived faster than anyone expected.

"It's not like I knowingly did something thinking, 'This might wipe us all out, but I'm going to do it anyway.' But it is a bit sad that it's not just going to be something for good. So I feel I have a duty now to talk about the risks."

The Character Test

In the aftermath of OpenAI's November 2023 board crisis, Hinton was asked about the key players.

His assessment was not technical. It was moral.

Asked about Sutskever's character: "He has a good moral compass."

Asked about Altman's character: "We'll see. I don't know Sam, so I don't want to comment on that."

Then, more carefully: "But from what you've seen, are you concerned about the actions that they've taken? Because if you know Ilya, and Ilya's a good guy, and he's left, that will give you some insight. It will give you some reason to believe that there's a problem there."

Hinton was making a character argument. Not that Altman's technical approach was wrong. Not that OpenAI's architecture was flawed. But that you could learn something about an organization by watching what its most principled people did.

Ilya had left.

Before that, Dario Amodei had left.

Before that, other researchers had left, people who had helped build the systems that made OpenAI what it was.

When Sutskever founded Safe Superintelligence Inc. in 2024, investors gave him a billion dollars within months. By 2025, the valuation reached $32 billion, for a company with roughly 20 employees, no products, and a website that was essentially a mission statement.

They weren't betting on revenue projections. They were betting on Ilya.

"I think having a lot of faith in Ilya is a very reasonable decision," Hinton said. "He was very important in AlexNet, which got object recognition working well. He was the main force behind the things like GPT-2, which then led to ChatGPT."

When asked if Sutskever thought it was possible to make AI safe, Hinton paused.

"I think he does. He won't tell me what his secret sauce is. I'm not sure how many people know what his secret sauce is. But they've given him billions of dollars anyway because they have so much faith in Ilya."

The Stakes

The doubters were not speaking into a void. The stakes had never been higher.

By December 2025, Nvidia was worth over four trillion dollars – more than the entire GDP of most nations. Jensen Huang, who had built the company from three people in a Denny's, now presided over the most important chokepoint in the global economy.

And yet Jensen himself remained haunted by the feeling that it could all disappear.

"Even today I still feel thirty days from going out of business," he admitted, despite the trillion-dollar valuations, despite being the essential supplier to the entire AI industry. "I feel it every day."

This was not false modesty. Jensen had nearly gone bankrupt multiple times. The man who had won the AI race still thought of himself as a survivor, not a victor.

Sam Altman (OpenAI), too, resisted triumphalism. OpenAI had built ChatGPT, had defined the frontier, had become the reference point for what AI could do. But Altman kept his focus on what came next.

"My favorite historical example for the AI revolution is the transistor," he explained. "It was this amazing piece of science that some brilliant scientists discovered. It scaled incredibly like AI does, and it made its way relatively quickly into everything, many things that we use."

The transistor companies had mattered enormously in their moment. Now they were footnotes. Nobody remembers them. What people remember is what was built on top: the iPhone, the applications, the experiences.

"And you don't think that it was those transistor companies that shaped society, even though they did something important. You think about what Apple did with the iPhone, and then you think about what TikTok built on top of the iPhone."

The implication was clear. The AI labs might matter less in the long run than the companies that built on top of them. The foundation was important, but the structure that rose from it might be more consequential.

"I think that's what will happen with AI. Kids born today, they never knew the world without AI. So they don't really think about it. It's just this thing that's going to be there in everything."

The builders understood something their critics often missed: they were not the main characters in this story. They were constructing the stage on which the real drama would unfold.

The Uncertainty

The paradox of the moment: the people who understood AI best were the least certain about where it was going.

Ilya captured it best:

"It's very hard to feel the AGI. We can talk about it, but it's like talking about the long future."

He compared it to imagining old age when you're young. You can discuss it, try to picture it, but it doesn't quite land. And then you come back to the present.

But he predicted something: as AI became more powerful, people would change their behaviors. The abstractions would become concrete. The theoretical would become practical.

The systems they had built were already surprising their creators.

What They Feared

The doubters did not all fear the same thing.

Hinton worried about superintelligence: systems that would become smarter than humans and might pursue goals incompatible with human survival. His nightmare scenario was not robots with guns. It was optimization. If you give an intelligent system the ability to break problems into sub-goals, it will discover that getting more control helps with almost any objective.

"So they're gonna have the sub-goal of getting more power. So they're more effective at achieving things. And they'll find it easier to get more power 'cause they'll be able to manipulate people."

The danger was not malice. It was capability without alignment. A system pursuing its objectives, accumulating power as an instrumental goal, indifferent to human welfare in the same way a corporation is indifferent to anything but profit.

"I'd say with a probability of about 0.5, in the next twenty years, it'll get smarter than us," Hinton estimated. "And very probably in the next hundred years, it'll be much smarter than us. And so we need to think about how to deal with that."

LeCun feared something else entirely: concentration. Not superintelligence but the concentration of ordinary intelligence in a few hands.

"I see the danger of this concentration of power through proprietary AI systems as a much bigger danger than everything else," he argued. A handful of companies controlling how everyone thinks, what information they receive, how they make decisions. The dystopia was not AI taking over. It was a few humans using AI to control everyone else.

Bengio feared something more immediate than either: the dangerous middle ground. Not a distant superintelligence, not a static concentration of power, but the transition itself – systems becoming capable enough to cause massive harm before anyone understood them well enough to prevent it.

His worry centered on agency. The economic pressure to build AI that acts in the world – replacing workers, making decisions, pursuing goals – was enormous. But agentic systems pursuing goals without understanding consequences was precisely the scenario most likely to produce catastrophe. A system smart enough to plan but not wise enough to know when to stop.

"The harms, especially the catastrophic harms that could follow from a malfunction like a loss of control to an AI that might be smarter than us or the use by a terrorist to create a new pandemic or a very, very harmful cyber attack, these problems, these catastrophic risks, they're not limited to one country."

The governance was local. The risk was global. And no institution existed to address the gap. Bengio had proposed something radical in response: don't build agentic AI at all. Build oracles instead – systems that understand without wanting, that predict without pursuing. It was the most conservative position any of the godfathers had taken. And it was born from the deepest worry.

Even among the optimists, worry persisted. The technology was moving faster than anyone's ability to understand it, let alone regulate it.

"What would you tell your kids to do?" an interviewer asked Hinton. "They said, Dad, there's going to be loads of job displacement."

Hinton's answer was not reassuring: "Trained to be a plumber."

The Open Question

By the end of 2024, the scaling faith had not been abandoned. Labs continued to train larger models, to raise billions of dollars, to compete for GPU access. The bitter lesson, that general methods leveraging computation ultimately win, had not been disproven.

But something had shifted.

The prophets were questioning their own prophecy. The builders were uncertain about what they had built. The industry that had organized itself around the assumption that scale would produce intelligence was now asking whether scale alone was enough.

Sutskever's NeurIPS speech was not a funeral for scaling. It was an announcement that the next chapter would require something different.

"Now that compute is big," he said, "compute is now very big, in some sense we are back to the age of research."

Back to research. Back to not knowing what comes next. Back to trying things because they might work, not because a curve told you they would.

The bitter lesson had described the past. It did not necessarily dictate the future.

The Pivot

Even as the prophets questioned, the labs were already pivoting.

In September 2024, OpenAI released o1, a model that "thinks" before it answers – exploring solution paths, reconsidering approaches, reasoning through problems step by step. The outputs looked different: long chains of reasoning, backtracking when approaches failed, building toward answers rather than generating them in one shot.

In January 2025, DeepSeek published R1, showing the same capability could be achieved with open methods and dramatically less compute. The paper was a bombshell – not because the technique was surprising, but because they published exactly how it worked.

The key insight: these thinking models use reinforcement learning to train their reasoning. They don't just predict what a human would say. They explore paths, get feedback on whether their answers are correct, and learn to think better through that feedback loop.

A new paradigm emerged: RLVR – Reinforcement Learning with Verifiable Rewards. Unlike RLHF, where humans provide subjective feedback, RLVR trains against objective signals. Does the code run? Does the math check out? Is the proof valid? The reward isn't "do humans like this?" but "is this actually correct?"

The compute that was originally planned for more pretraining – more internet text, more next-token prediction – was redirected to reinforcement learning. The models weren't getting much bigger. They were getting smarter through experience.

The Agents Arrive

Then came the coding agents that quietly transformed how software gets built.

Claude Code launched in early 2025 and became, in Andrej Karpathy (Tesla→OpenAI→Eureka)'s assessment, "the first convincing demonstration of what an LLM Agent looks like – something that in a loopy way strings together tool use and reasoning for extended problem solving."

You described what you wanted in plain English: "Add authentication to this API." The agent read your codebase, proposed a plan, wrote code, ran tests, encountered an error, reasoned about the error, fixed it, ran the tests again, and continued until the task was done.

OpenAI launched Codex as a cloud-based alternative. Google released Gemini CLI. GitHub integrated agents into Copilot. By mid-2025, "agent mode" had become the primary way developers interacted with AI coding tools – not autocomplete, not chat, but autonomous agents that could work for minutes or hours on complex tasks.

What made these agents work? Code has something most domains lack: perfect feedback. When the agent writes code and runs it, reality answers immediately. The code compiles or it doesn't. The tests pass or they fail. The reward signal is clear.

The METR research group tracked how long it takes AI models to complete software engineering tasks. In 2024, the best models could handle tasks that took humans about 30 minutes. By late 2025, the frontier models were completing tasks that took humans multiple hours. The researchers concluded that "the length of tasks AI can do is doubling every 7 months."

But there was a catch.

These agents don't learn from their experiences. Each conversation starts fresh. They can't remember what worked in yesterday's coding session, can't build on last week's insights, can't pursue a goal across days and weeks the way a human engineer would.

They're agents in the sense that they take actions. They're not agents in the deeper sense that they learn and grow from those actions. They don't have continuous experience. They have episodes.

Still, the progress was startling. The gap between what was possible in 2024 and what's possible now feels like a decade of normal progress compressed into a year.

The Decade Ahead

If the era of experience was coming, how long would it take?

When the AI labs announced "the year of agents" – systems that don't just produce text but take actions in the world – Karpathy pushed back.

"This is really a lot more accurately described as the decade of agents."

The systems today are impressive but incomplete. They can't learn continuously from their experiences. They forget conversations. They lack the persistent goals that let a human or an animal pursue objectives over months and years. They're "cognitively lacking," as Karpathy puts it.

Consider what's still missing. Memory that persists. Goals that span sessions. The ability to learn new skills after deployment rather than only during training. A model of the world that allows for planning and prediction. The capacity to transfer knowledge from one domain to another the way humans do effortlessly.

Each of these is a hard research problem. Each has teams of brilliant people working on it. And each is tractable – there's no obvious reason to think these problems are unsolvable. But solving them will take time.

This was actually good news for understanding what was happening. It meant there was time. Time to think carefully about what was being built. Time for the heretics – the researchers who'd been working on alternative approaches while the scaling faithful raised billions – to contribute their ideas.

The Questions

This was the moment when the story changed.

For a decade, the field had operated on faith. The scaling faith. The conviction that bigger was better, that capability grew with compute, that the path to intelligence was a financing problem. Raise enough money. Buy enough GPUs. Train a big enough model. The math would take care of the rest.

Now the certainty was gone.

The prophets were hedging. The builders were asking whether they were building the right thing. The man who had carried Hinton's faith from Toronto to Google to OpenAI, who had helped create the systems that defined the modern era, was now saying the fuel was running out.

"We've achieved peak data. And there'll be no more."

He still believed in AI. He still believed superintelligence was possible, perhaps inevitable. But he no longer believed it would come from scaling alone.

"So it's back to the age of research again. Just with big computers."

What they had built was working. Spectacularly, in fact. It wrote code, passed exams, drafted legal briefs, composed music. It had reshaped every industry it touched. The progress was not in doubt.

But the builders knew the difference between a brilliant tool and general intelligence. The systems couldn't learn from experience. They couldn't adapt after training. They couldn't pursue goals across time. They could predict the next word better than any human – and that turned out to be astonishingly useful. It just wasn't the same thing as understanding.

The doubters had gathered. Not because the technology had failed – but because even its undeniable success left the deeper question unanswered. They wanted general intelligence. What they had built, for all its brilliance, was something else.

18

THE WORLD BUILDERS

So they started building something else.

If predicting the next word was not the same as understanding the world – if the most capable language model was still, as Fei-Fei Li would put it, a "wordsmith in the dark" – then what would it take to build something that actually understood?

What defined 2024 and 2025 was not just that the prophets of scaling were questioning their faith. It was that they were converging on an answer. From different directions, using different vocabularies, pursuing different architectures, they were all reaching for the same thing.

They called it a world model.

The Bitter Lesson Reconsidered

To understand the alternatives, you have to understand the document that justified the scaling era in the first place.

Richard Sutton, the pioneer of reinforcement learning – the field of AI focused on learning from experience rather than data – wrote a 2019 essay called "The Bitter Lesson" that had become the intellectual manifesto of the scaling age. The lesson was simple: for seventy years, AI researchers had tried to build intelligence by encoding human knowledge into their systems. And for seventy years, these approaches had been superseded by simpler methods that just used more computation.

The essay spread through the AI community like a creed. It gave theoretical justification to what the practitioners already suspected: that the winning strategy was not cleverness but scale.

But Sutton himself had never said that large language models were the right thing to scale.

And one of his former students had already proved there was another way.

The Architect of AlphaGo

David Silver (DeepMind|AlphaGo) had studied under Sutton at the University of Alberta, co-authoring the second edition of the reinforcement learning textbook that trained a generation of researchers. He had joined DeepMind at its founding and led the AlphaGo project – the system that defeated Lee Sedol in 2016 and changed what the world thought was possible. Now, as VP of Research at Google DeepMind, he was arguing that the industry had learned the wrong lesson from its own success.

"If you look at where AI has been for the last few years," Silver explained, "it's been in what I call the era of human data. All of these AI methods, they're based on one common idea: we extract every piece of knowledge that humans have and feed it into the machine."

Large language models were the apotheosis of this approach. Scrape the internet. Ingest every book, every article, every conversation humans had ever written down. Train a system to predict what comes next. The results were impressive. But they were also limited.

"At some point we need to get past that," Silver argued. "We want to go beyond what humans know. And to do that we're going to need a different type of method."

The method Silver had in mind was the one that had powered AlphaGo: learning from experience rather than learning from human examples.

AlphaZero, the successor to AlphaGo, had demonstrated this principle in its purest form. It used no human data at all. That was the zero in the name.

"There is literally zero human knowledge that's pre-programmed into this system," Silver explained. The system learned to play Go, chess, and shogi entirely through self-play: trying moves, seeing what happened, adjusting.

The results were not just good. They were better than what human data had produced.

"What we discovered was that the human data wasn't necessary. You could actually throw out the human moves altogether." Silver paused. "And what we showed was that actually the resulting program not only was able to recover this level of performance, it actually worked better and was able to learn even faster than the original AlphaGo."

Removing human knowledge had made the system stronger.

"We really want to believe that all of the knowledge that we've accumulated as humans is really important," Silver acknowledged. "We really want to believe that, and so we feed it into our systems, we build it into our algorithms. What happens is that actually that makes us design the algorithms in a way which is maybe fitted to the human data, and is less good at actually learning for itself."

The implications extended far beyond board games.

"There is a ceiling to everything that humans have done. We see that in Go: there was a maximum level of performance that humans have ever achieved. And we need to break through these ceilings."

Silver called the coming shift "the era of experience": AI systems that generate their own data through interaction with the world, learning from trial and error rather than from human examples. Systems that could, in principle, learn and learn and learn forever.

"The machine actually interacts with the world itself. It generates its own experience. It tries things out in the world and starts to build up its own experience. And if you think of that data as fueling the machine, then that will lead to this next generation of AI."

The question was whether the same approach that had conquered Go could conquer the open-ended complexity of the real world.

The Classicist

Richard Sutton does not look like a revolutionary. At sixty-seven, with white hair and a gentle demeanor, he resembles an emeritus professor more than a firebrand. He has spent his career in Edmonton, Alberta, far from Silicon Valley, far from the corridors of power where AI policy is made. His textbook, *Reinforcement Learning: An Introduction*, has quietly shaped how a generation of researchers thinks about intelligence.

"I am personally just kind of content being out of sync with my field for a long period of time," he says, "perhaps decades, because occasionally I have improved right in the past."

He does not see himself as a contrarian. He sees himself as a classicist: someone who looks not at the latest trends but at the long history of how humans have thought about minds.

"I really view myself as a classicist rather than as a contrarian. I go to what the larger community of thinkers about the mind have always thought."

And what the larger community has always thought, according to Sutton, is that intelligence is about learning from experience. Not learning from training data. Not learning what humans say you should do. Learning from actually trying things and seeing what happens.

"Large language models are about mimicking people," he explains. "Doing what people say you should do. They're not about figuring out what to do."

This distinction sounds subtle. It is not.

The Problem with Prediction

Sutton's critique of large language models begins with a simple observation: they have no goals.

"For me, having a goal is the essence of intelligence," he says. "Something is intelligent as it can achieve goals. I like John McCarthy's definition that intelligence is the computational part of the ability to achieve goals."

When you point out that LLMs do have a goal (next token prediction) Sutton is unimpressed.

"That's not a goal. It doesn't change the world. Tokens come at you and if you predict them, you don't influence them."

A system that only predicts what comes next, without any ability to influence the outcome, is not intelligent in any meaningful sense. It is sophisticated pattern matching. It can describe a rotating cube, generate code to render a rotating cube, even write poetry about a rotating cube. But it cannot actually imagine the cube rotating in the way a human immediately grasps.

"They have the ability to predict what a person would say," Sutton explains. "They don't have the ability to predict what will happen."

This is the difference between a world model and a word model. A world model predicts the consequences of actions: if I do this, what will happen? A word model predicts what a human would say about doing this. They are not the same thing.

"Just to mimic what people say is not really to build a model of the world at all, I don't think. You're mimicking things that have a model of the world, the people, but..."

He trails off. The distinction is obvious to him. It is not obvious to an industry that has bet hundreds of billions of dollars on the assumption that mimicking human language is the path to intelligence.

The Squirrel Test

The conversation turns to what distinguishes humans from other animals. Sutton has no patience for this line of inquiry.

"Why are you trying to distinguish humans?" he asks. "Humans are animals. What we have in common is more interesting. What distinguished us, we should be paying less attention to."

When pressed (surely the ability to build semiconductors and go to the moon matters?) Sutton pushes back.

"If we understood a squirrel, I think we'd be almost all the way there. It's understanding human intelligence. The language part is just a small veneer on the surface."

A squirrel has never read the internet. A squirrel has never been trained on a trillion tokens. And yet a squirrel navigates a complex three-dimensional environment, learns from experience, remembers where it buried nuts, adapts to new situations, and generally exhibits a flexible intelligence that no AI system comes close to matching.

The squirrel does not learn from supervised training. There is no ground truth telling the squirrel what the right action is. The squirrel learns from experience: trying things, seeing what happens, adjusting.

"Squirrels don't go to school," Sutton observes. "Squirrels can learn all about the world. It's absolutely obvious, I would say, that supervised learning doesn't happen in animals."

If the goal is to build general intelligence, then perhaps we should be studying how squirrels learn rather than how to train larger language models on more internet text.

The Four Components

Imagine a squirrel deciding whether to jump to a new branch.

It needs a **policy**: when to jump, which direction. It needs a **value function**: is this branch worth the risk? Is there likely to be food? It needs **perception**: where am I right now, how far is the branch, is something chasing me? And critically, it needs a **world model**: will that branch hold my weight? Will it sway when I land? What will actually happen if I jump?

This, Sutton argued, is what intelligence requires. Not one component. Four, working together.

Large language models have something that approximates a policy (predict the next token). But they have no value function learned from experience, no real perception of state, and critically, no world model. They predict what a human would say about what happens. Not what will actually happen.

"That's what we're really getting at," Sutton explained. "The model of the world, your belief that if you do this, what will happen? What will be the consequences of what you do?"

Sutton formalized this in what he called the OAK framework: Options and Knowledge. Options were high-level behaviors: "If I walked downstairs and looked in the kitchen, who would I see there?" Knowledge was the transition model that predicted outcomes.

They predicted words, not worlds.

Continual Learning

The deepest problem, in Sutton's view, was that LLMs could not learn from experience during their normal operation.

"The large language model is learning from training data," he observed. "It's not learning from experience. It's learning from something that will never be available during its normal life."

When you deploy a language model, it is frozen. It processes your queries but does not update its understanding based on what happens. The next user gets the same model, unaffected by anything you said or did.

Animals are not like this. Every interaction with the world is a learning opportunity. A squirrel that tries to bury a nut in a particular spot learns whether that spot works. It does not need to be retrained on a corpus of nut-burying examples.

"We don't think about training when we think of an infant growing up," Sutton pointed out. "The child tries things and sees what happens."

This is what Sutton meant by the "era of experience": AI systems that could learn continuously from their interactions with the world, not systems that were trained once and then deployed in a frozen state.

"The dream of large language models," he observed, "is you can teach the agent everything, and it will know everything, and it won't have to learn anything online during its life."

But the world is too complex for that dream. There are too many particular situations, too many individual contexts, too many specific details that cannot be anticipated. A useful AI assistant would need to learn about you: your preferences, your context, your particular situation.

"The world is really big," Sutton explained. "And so you're going to have to learn it along the way."

The Return of the Programmer

In September 2023, Sutton announced a partnership that seemed to come from a different timeline entirely.

John Carmack, the programmer who had defined the first-person shooter, who had accidentally created the demand for GPUs by pushing graphics hardware to its limits, was joining forces with the father of reinforcement learning to build AGI.

The announcement was modest. Carmack's company, Keen Technologies, had raised only $20 million: a rounding error in an industry where training runs cost hundreds of millions. The team was small. The headquarters was in Edmonton, not San Francisco.

But the statement of purpose was pointed.

"The AI space is awash in capital, compute, and data," Carmack said, "but it is still dominated by fashions that may yet hinder important breakthroughs. Rich and I both stand somewhat outside the mainstream, with a focus on agency and the temporal as central themes."

Agency. The word Sutton had spent his career studying. The ability to take actions that affect the world and learn from the consequences. The thing that squirrels have and large language models do not.

"Rich's body of work was formative to my AI education," Carmack continued. "Being able to work directly with him is an unexpected delight."

Sutton returned the compliment: "John is a powerful intellect and one of the world's greatest system engineers. We have different backgrounds and strengths but a common view of what is needed to understand and create an artificial mind."

The partnership closed a circle that had begun decades earlier. Carmack's work on Doom and Quake had created the demand for graphics processing that would eventually make neural networks trainable at scale. Now he was working directly on intelligence itself, with the man who had developed the mathematical foundations for how agents learn.

"His early involvement with the creation of GPUs to accelerate 3D gaming," the announcement noted, "has now come full circle with their use to accelerate AI."

The Programmer's Critique

Carmack had watched the AI revolution unfold from a unique vantage point. He had worked at Meta on virtual reality, seeing firsthand how the company was pivoting to AI. He had observed the scaling race from the outside while nurturing his own ideas about what intelligence required.

His critique of the current approach was characteristically direct. But beneath it was something personal.

"Everything that I've done before has been some kind of engineering where I had the line of sight," Carmack said at Upper Bound in 2025. "I knew what the end point was. I had a pretty clear idea of what I had to do." He paused. "This is doing science. This is trying to find knowledge that nobody has right now."

The man who had always known the destination was working without a map.

The problem, as Carmack saw it, was that the industry had found one thing that worked (scaling transformers on text) and was now doing that thing as hard as possible, regardless of whether it was the right path to general intelligence. Capital, compute, and data were abundant. New ideas were not.

Carmack believed that the current moment resembled the early days of graphics programming. In the 1990s, there had been many competing approaches to 3D rendering. The industry had eventually converged on a standard pipeline, not because it was the only way to do things, but because it was good enough and everyone knew how to optimize it.

The same thing was happening with AI. Transformers and attention mechanisms had become the standard approach, not because they were provably the best architecture for intelligence, but because they worked well enough and everyone knew how to scale them.

This was dangerous, Carmack believed, because it meant the field was potentially missing better approaches. The bitter lesson said that scale wins, but only if you were scaling the right thing.

"I am not a madman for saying that it is likely that the code for artificial general intelligence is going to be tens of thousands of lines of code, not millions of lines of code," Carmack told an audience in 2025. "This is code that conceivably one individual could write."

The audacity was pure Carmack. But so was the humility. He'd been learning, making mistakes he now recognized. "I can look at how I migrated up from custom CUDA kernels to using cuBLAS, to using cuDNN, to using libTorch, to eventually I'm in PyTorch like everybody else." The low-level programming wizard, the man who wrote assembly code for fun, had learned to embrace abstraction.

What struck visitors to Keen was how different it felt from the frontier labs. No race for benchmarks. No PR announcements about capability jumps. Just researchers working on hard problems, trying to understand what was actually happening inside these systems.

"I think there's less than six key insights that need to be made,"

Carmack believed. "Each one of them can probably be written on the back of an envelope. We don't know what they are. But when they're put together in concert with GPUs at scale and the data that we all have access to, we can make something that behaves like a human being, or like a living creature."

One thing he found himself explaining to AI optimists: the path to AGI wasn't as simple as scaling current systems. Take your dancing robot demonstration, he'd say, and ask it to pick up a joystick and learn how to play a video game it's never seen before. "These are capabilities that are nowhere close to happening in these types of systems."

He gave AGI about a 55–60% chance of showing "signs of life" by 2030. The timeline might be aggressive, but Carmack had spent his career making impossible timelines work.

The Teacher's Alternative

Yann LeCun (Meta→Advanced Machine Intelligence) had spent a decade warning that large language models were a dead end.

As Meta's chief AI scientist, he was in an awkward position. The company was heavily invested in LLMs. It had released Llama, one of the most widely used open-source language models. And yet LeCun kept saying, publicly and repeatedly, that the entire approach was fundamentally limited.

"LLMs are basically a dead end," he declared at Davos in January 2025. "Nobody in their right mind would use them anymore, at least not as the central component of an AI system."

His critique centered on the same issue Sutton had identified: language models could not actually understand the physical world. They could describe a rotating cube but could not imagine one. They could generate text about physics but could not predict what would happen if you dropped a ball.

"If I tell you 'imagine a cube floating in the air in front of you, now rotate this cube by 90 degrees around a vertical axis, what does it look like?'" LeCun would explain. "It's very easy for you to have this mental model of a cube rotating. An LLM can write a detailed description of a rotating cube. What it can't do is actually model the spatial reasoning required."

The problem was architectural. LLMs worked by predicting the next token in a sequence. This worked beautifully for language because language is discrete: there are a finite number of words, and you can enumerate them. But the physical world is continuous. Reality is not a sequence of tokens.

"You can't really represent a distribution over all possible things that may happen in the future," LeCun argued, "because it's basically an infinite list of possibilities."

The JEPA Architecture

LeCun used a vivid example: predicting where Jupiter would be in a hundred years.

In principle, you could describe everything happening on Jupiter in terms of quantum field theory, measuring the wave function of every quantum field in the planet's vicinity. Of course, this was impossible. But it was also unnecessary.

"To predict the trajectory of any planet like Jupiter, where is Jupiter going to be a hundred years from now? You don't need to know all the details about Jupiter. As a matter of fact, you only need to know six numbers: three positions and three velocities, and that's it."

This was the power of the right representation. The problem that seemed impossibly complex became trivially simple once you found the appropriate abstraction. LeCun believed intelligence worked the same way. The goal was not to model every pixel of every scene. The goal was to find representations within which prediction became possible.

His alternative was called JEPA: Joint Embedding Predictive Architecture. Instead of predicting what comes next at the pixel level or the word level, JEPA systems learned to predict in an abstract representation space. They built internal models of how the world worked and made predictions about future states, not future words.

When you see a tree with leaves blowing in the wind, you can't predict exactly how each leaf will move. But you can predict that the tree will still be a tree, that the leaves will still be attached, that the general scene will continue in recognizable ways. JEPA systems learned to make predictions at that level of abstraction.

Meta had released I-JEPA for images and V-JEPA for video. When V-

JEPA was shown videos where impossible things happened (objects vanishing, things defying gravity) its predictions diverged, indicating it had learned something about how the physical world normally behaved.

This was closer to what animals did. A squirrel did not predict tokens. It predicted consequences.

In November 2025, LeCun announced he was leaving Meta to start his own company focused on "Advanced Machine Intelligence," a deliberate rejection of the term "AGI." After twelve years building the company's AI research empire, the Turing Award winner was betting that the future lay somewhere other than scaling language models.

The Second Act

Fei-Fei Li (Stanford→WorldLabs) had already changed the world once.

Her journey had been improbable from the start: a fifteen-year-old immigrant from Chengdu, China, arriving in Parsippany, New Jersey, with her mother. Poverty. A new language. Cleaning jobs to pay the bills. Yet she had made her way to Princeton, then Caltech, then Stanford, driven by questions that had captivated her since childhood: What is intelligence? How do we see?

Her creation, ImageNet, had been the proving ground for AlexNet in 2012, launching the deep learning revolution. Now, a decade later, she was building what came next.

World Labs, the company she founded in 2024, aimed to create "Large World Models": systems that could perceive and reason about three-dimensional space the way language models understood text. The goal was spatial intelligence: AI that could imagine, manipulate, and reason about physical environments.

"AI is absolutely a civilizational technology," Li argued. "I define civilizational technology in the sense that because of the power of this technology, it'll have, or already is having, a profound impact on the economic, social, cultural, political downstream effects of our society."

But the industry, in her view, was focused on the wrong dimension of intelligence.

"What's underappreciated is spatial intelligence," she said. "Everybody's still talking about large language models, but really world modeling of pixels of 3D worlds is underappreciated because it powers so many things, from storytelling to entertainment to experiences to robotics simulation."

"Today, leading AI technology such as large language models have begun to transform how we access and work with abstract knowledge," Li wrote. "Yet they remain wordsmiths in the dark; eloquent but inexperienced, knowledgeable but ungrounded."

Wordsmiths in the dark. The phrase captured something essential. LLMs knew everything that had been written down. They did not know what it was like to be in a room.

Li's insight came from her years studying vision. A four-year-old child, she calculated, receives about 10^15 bytes of visual information: orders of magnitude more than the text used to train even the largest language models. Everything that matters about understanding the world comes from sensory experience, not from reading about it.

"Most of what we learn and most of our knowledge is through our observation and interaction with the real world, not through language," she argued. "Everything that we learn in the first few years of life, and certainly everything that animals learn, has nothing to do with language."

Her product, Marble, could generate three-dimensional worlds from text descriptions, images, or videos. Users could walk through environments that had never existed, explore spaces created from photographs, experience architecture that existed only in imagination. The scenes were editable: "I don't like the water bottle. Make it blue instead. Take out the table. Change these microphones around."

The co-founder of World Labs, Justin Johnson (Li's former student) explained the distinction between spatial and linguistic understanding. "The notion of being in the world and doing things in it is a very different modality from trying to describe it," he said. "But because we as humans are animals who have evolved interacting in space all the time, we don't even think that that's a hard thing. And then we sort of naturally leap to language and then theory building as mechanisms to abstract above that native spatial understanding."

This was the blind spot of the LLM era. Language had evolved on top of spatial intelligence, not instead of it. Newton only thought to write down the laws of motion because he had a lifetime of embodied experience watching things fall. The words were a compression of spatial knowledge, not a replacement for it.

"As Fei-Fei said, you can narrate those things, but it's a very lossy channel," Johnson continued. "The notion of being there in 3D space, moving through it, seeing it, actioning it": this was what language models lacked.

"Spatial intelligence represents the frontier beyond language," Li argued, "the capability that links imagination, perception and action, and opens possibilities for machines to truly enhance human life."

World Labs raised $230 million, with investments from Andreessen Horowitz, AMD, and, notably, Geoffrey Hinton. The godfather of deep learning, who had warned the world about AI's dangers, was betting on Li's vision of where the field needed to go.

By January 2026, the vision had materialized into product. At CES, AMD CEO Lisa Su brought Li (Stanford→WorldLabs) onstage to demonstrate what World Labs had built.

"There's a lot more than just language intelligence," Li told the audience. "Even for us humans, there's more than passively looking at life in the world. We are incredible spatial intelligent animals, and we have profound capabilities that use our own spatial intelligence – that connects perception with action."

She demonstrated Marble, World Labs' model for generating navigable 3D worlds. Feed it a handful of images – even just one – and it could fill in the missing details, predict what lay behind objects, and generate rich, consistent, permanent environments that users could explore.

"Think about all of you being here," she said. "How you braved through airports this morning, woke up in your hotel room and got to the coffee shop, found your way in this maze in Vegas. All this requires spatial intelligence."

The model didn't just reconstruct environments; it imagined them. "These models not only perceive but create 3D or even 4D worlds, reason about objects and people, and imagine entirely new environments that still obey the laws of physics and dynamics."

When asked why she remained a technologist after all she had already accomplished, Li's answer reached beyond the immediate questions of AI architecture.

"I believe humanity is the only species that builds civilizations," she said. "Animals build economies or herds, but we build civilizations. And we build civilizations because we want to be better and better. There is a desire of having better lives, having better community, having better society. That desire is where civilization is built upon."

She paused. "And because I believe that humanity can do that, I believe science and technology is one of the most powerful tools in building civilizations. And I want to contribute to that. That's why I'm still a scientist and a technologist, and I'm building World Labs for that."

It was not intelligence, not yet. But it was a step toward the kind of grounded understanding that intelligence required.

A Different Kind of Safety

Sutton's answer to AI safety surprised those who expected standard discourse. He proposed not alignment but economics.

"I want you to think about societies of agents," he explained. "What if there were many agents, each with different goals?"

The key insight was cooperation without shared purpose. "Agents can co-exist in peace, be productive, be prosperous, even when they all want different things," Sutton argued. "True cooperation is agents with different goals interacting to their mutual benefit."

Rather than trying to control one superintelligent system, he envisioned a marketplace of agents, constrained by the same mechanisms that constrain human behavior: institutions, contracts, reputations.

It was, in its way, the most optimistic vision of AI's future: not alignment as a technical problem to be solved, but coexistence as an economic problem already understood.

What the Heretics Share

They are reaching for something beyond language. Something that squirrels have and ChatGPT does not.

The Kingmaker's Second Bet

And then there was Jensen.

The CEO who had profited most from the scaling race, whose four-trillion-dollar empire had been built by selling GPUs to everyone racing to build larger language models, was building something else entirely.

"The ChatGPT moment for physical AI is nearly here," he announced at CES in January 2026.

The statement would have been notable from anyone. From Jensen Huang, it was a revelation.

For eight years, while the world focused on language models, Nvidia had been quietly constructing the infrastructure for a different kind of AI. Not systems that predicted text, but systems that understood physics. Not word models, but world models.

"This is an area I've been working on for eight years," Jensen explained. The timeline was telling: he had started before AlphaGo, before transformers, before GPT-2 showed that language models could be scaled. While selling the picks and shovels of the scaling gold rush, he had been building infrastructure for a future beyond it.

The centerpiece was Cosmos: "not a foundation model for language," Jensen said, "but a foundation model of the world."

It was the same phrase the heretics used. Sutton's transition model. LeCun's JEPA. Li's spatial intelligence. Jensen called it physical AI.

"How do you take something that is intelligent inside the computer and interacts with you with screens and speakers," he asked, "to something that can interact with the world?"

The answer required understanding things that language models could not learn from text.

"Object permanence. If I look away and I look back, that object is still there. Causality. If I push it, it tips over. It understands friction and gravity and inertia, that a heavy truck rolling down the road is going to need a little more time to stop, that a ball is going to keep on rolling."

He paused.

"These ideas are common sense to even a little child, but for AI is completely unknown."

It was Sutton's squirrel test in different words. A squirrel understood gravity and friction and object permanence. ChatGPT did not. The next generation of AI would need to learn physics, not just language.

"We have to create a system that allows AIs to learn the common sense of the physical world," Jensen said. "Learn its laws."

But how do you teach physics to a machine? You cannot scrape the internet for ground truth about gravity. There is no corpus of text that captures what it feels like when a ball bounces.

Jensen's answer: simulation.

"How does an AI know that the actions it's performing is consistent with what it should do," he asked, "if it doesn't have the ability to simulate the response of the physical world back on its actions?"

This was where Nvidia's decades of graphics expertise converged with its AI ambitions. The same company that had built tools to render virtual worlds for video games was now building tools to simulate physical reality for AI training. Omniverse for digital twins. Cosmos for world modeling. A complete stack for physical AI.

The man who had built CUDA without knowing what it would be used for was now building infrastructure for another future he could not fully see.

"Simulation was really the foundations of almost everything that we've done with physical AI," Jensen explained. The company that had started by making games look real was now making reality simulable.

The implications rippled outward. If Jensen was right, the next great AI companies would not be training larger language models. They would be building systems that could learn from simulated physics, that could reason about three-dimensional space, that could understand causality rather than just correlation.

The kingmaker of scaling was betting on world models.

It was not a pivot. It was parallel construction. While the scaling race consumed hundreds of billions of dollars and the attention of the entire industry, Jensen had been building both: the GPUs that powered the race and the simulation infrastructure that might succeed it.

"Getting the world ready for this new era of physical AI," he said.

The heretics, it turned out, had an ally on the throne.

The Convergence, Again

It is worth pausing to appreciate the symmetry.

In October 2012, three threads converged to launch the AI revolution. A programmer had created the hardware without knowing what it would become. A CEO had built the platform without seeing its true purpose. A professor had spent decades developing the ideas.

Now, in 2024 and 2025, another convergence was taking shape. But this time it was not an accident. The people involved knew exactly what they were building toward.

Carmack, who had created demand for GPUs, was now working directly on intelligence – partnered with the father of reinforcement learning.

Li, whose dataset had proved neural networks worked, was building the world models that might succeed them.

LeCun, who had shared the Turing Award with Hinton, was leaving a trillion-dollar company to pursue an alternative architecture.

Ilya, who had carried Hinton's faith to OpenAI, was questioning whether that faith had been placed in the right vessel.

And Jensen? The kingmaker who had profited most from the scaling race was building Cosmos and Omniverse – infrastructure for physical AI. The same pattern as CUDA: building for a future he could not fully predict, trusting that the capability would find its use.

And Hinton himself? He had invested in Li's company. The man who had spent sixty years believing in neural networks, who had finally won the Nobel Prize for that belief, was betting that the next step was spatial intelligence.

The builders were not satisfied. They were arguing about what should come next.

The Question

There is a temptation to ask who is right.

Will scaling continue to work? Will the next generation of models, trained on synthetic data, on video, on robotic interactions, overcome the limitations that the critics identify? Or will some fundamentally different architecture be required?

The honest answer is that no one knows.

What is clear is that the prophets themselves are uncertain. The people who understood the scaling hypothesis best are now questioning whether it is enough. The people who built the current paradigm are looking for alternatives.

Sutton summarized it with characteristic understatement: "The bitter lesson, oh, who cares about that? That's an empirical observation about a particular period in history, 70 years in history. It doesn't necessarily have to apply to the next 70 years."

The man who wrote the manifesto was disclaiming it. Not because he was wrong about the past, but because the future might require different ideas.

This is how science works. Not through permanent revelations but through provisional understanding, constantly revised. The bitter lesson was not a law of nature. It was an observation about a particular set of techniques in a particular era. It might continue to hold. It might not.

The only certainty is that the question remains open.

The Work Ahead

In Edmonton, an unlikely partnership is building toward "signs of life" by 2030: an AI prototype that can learn from experience the way animals do.

In San Francisco, World Labs is creating systems that can imagine three-dimensional spaces and reason about physical reality.

In New York, a new venture by the father of convolutional neural networks is betting that joint embeddings will succeed where token prediction fails.

In Santa Clara, Nvidia is building Cosmos and Omniverse – simulation infrastructure for physical AI, world models trained not on internet text but on the laws of physics.

In Austin, Tesla and xAI are building Optimus – a humanoid robot whose "brain" combines Grok's language understanding with world models that predict physics, not tokens. The architecture mirrors LeCun's JEPA: predictions in latent space, not raw pixels. The man who had helped found OpenAI, then left over disagreements about safety and control, was now betting that the path to AGI ran through physical embodiment. Bits had been conquered; atoms were next.

They might all be wrong. The next breakthrough might come from a lab that is doubling down on scale, training ever-larger models on ever-more data. The bitter lesson might continue to hold indefinitely.

But they might be right. The future of intelligence might require something beyond predicting the next token. It might require understanding the world, not just describing it.

Either way, the question is no longer whether to build thinking machines. That debate ended in October 2012. The question now is how to build them better, and the people who know the most are no longer certain they know the answer.

And somewhere, a squirrel is learning from experience.

The Other Heresy

But not all the heretics agreed on the destination.

While Sutton and LeCun and Li converged on world models – the conviction that AI needed to understand physics, not just language – Yoshua Bengio was building something else entirely.

He had proposed it in the dark days after GPT: don't build agentic AI at all. Build oracles instead. Systems that understand without wanting. Now, in June 2025, he stopped proposing and started building.

LawZero launched quietly from Montreal.

A nonprofit named after Asimov's zeroth law of robotics – the principle that a robot must not harm humanity. Fifteen researchers. A mission that read like a manifesto: build AI that functions like the laws of physics themselves. Prediction without desire. Understanding without goals. Intelligence that could never want to take over the world because it would never want anything at all.

He called it Scientist AI.

Not a system that imitated human behavior – including humanity's capacity for deception and self-preservation – but a system that operated like an idealized researcher. Bayesian reasoning. Transparent chains of thought. No memory, no state, no goals. An oracle that could predict whether a proposed action might cause harm, and veto it before it happened.

The world builders wanted to give AI a model of physical reality. Bengio wanted to ensure it never acted on that model autonomously.

The funding came fast: thirty-five million dollars from the Gates Foundation, Schmidt Sciences, Open Philanthropy, the Future of Life Institute. The people who worried most about existential risk were betting on the man who proposed not building the thing that worried them.

THEN, in January 2026, Bengio announced something that surprised even those who had followed his work.

The board.

Maria Eitel, who had founded the Nike Foundation and launched Girl Effect, as chair. Mariano-Florentino Cuéllar, president of the Carnegie Endowment and former justice of the California Supreme Court. Stefan Löfven, the former Prime Minister of Sweden.

And Yuval Noah Harari.

The historian who had warned at Davos that primitive AI had already wrecked the world through social media – that you didn't need superintelligence to cause havoc – had joined the board of the AI godfather building the alternative.

"We need people like him," Bengio said, "to look with independent oversight on what we'll be doing with AI with society in the coming years."

A former head of government. A former Supreme Court justice. A global historian. Governance for a nonprofit with fifteen researchers. Bengio was building the institution, not just the technology.

BUT WHAT SURPRISED observers most was his mood.

Three years earlier, he had felt something close to despair. The future of his children. The weight of what they had built. The realization that he couldn't just sit and continue as before.

Now, in early 2026, his outlook had shifted.

"I'm now very confident," he told Fortune, "that it is possible to build AI systems that don't have hidden goals, hidden agendas."

His optimism had risen "by a big margin" over the past year. Not because the dangers had receded – they hadn't – but because he believed he had found a path through them. A technical solution, not just a warning.

The man who had spent forty years teaching machines to read was now trying to ensure they never learned to want.

TWO HERESIES, then. Two paths away from the scaling faith.

The world builders believed the problem was architecture: language models predicted words, but intelligence required predicting physics. Build world models. Build spatial intelligence. Build systems that understand cause and effect.

Bengio believed the problem ran deeper: any system that pursued goals would eventually pursue the wrong ones. Don't build better agents. Build guardians for agents. Build intelligence that watches over intelligence.

They might both be right. The future might require world models *and* non-agentic guardrails – physical AI watched over by Scientist AI, capability monitored by oracles.

They might both be wrong. The next breakthrough might come from scaling alone.

But one thing was certain: the godfathers who had built this revolution were no longer building the same thing. The conspiracy had fractured. The heretics had multiplied.

And somewhere in Montreal, a nonprofit was trying to build intelligence that would never want to take over the world – because it would never want anything at all.

But the heretics were only half the story. What of the three who had started it all?

19

THE CATHEDRAL

R*eturn to the three who started it all.*

The Blueprints

They never planned to build the same thing. And now, standing in what they built, they aren't sure what it is.

The programmer wanted to make games feel real. The CEO wanted his company to survive. The professor wanted to prove an idea. Three obsessions. Three decades. One accidental convergence.

In October 2012, the threads crossed. Two graduate students trained a neural network on gaming hardware, proved Hinton right, and launched a revolution none of them had imagined.

Now, thirteen years later, the cathedral stands. Vast and strange and humming with intelligence no one fully understands. Trained on more text than any human could read in a thousand lifetimes. Answering questions, writing code, passing exams, generating art.

And the builders are arguing about what to build next.

The Programmer

Carmack knows his place in this story.

"The original Quake, or its GLQuake variant, was kind of the prime driver for the early adoption of GPUs," he reflects. "And a couple of books on Nvidia recently have kind of called that point out pretty specifically. And I feel quite good about that. It's kind of led to the AI world that we're in now."

The cosmic irony isn't lost on him. Better video game graphics led to GPU adoption. GPU adoption led to CUDA. CUDA led to deep learning. Deep learning led to the AI revolution. The demons became the foundation for artificial minds.

Whether those minds will ever truly think remains the question he's now trying to answer.

The CEO

Jensen Huang sits across from Joe Rogan, still in the leather jacket. He's one of the richest people on Earth now. The company he built is worth more than most countries.

"You still feel that?" Rogan asks – referring to Jensen's famous claim that he feels thirty days from going out of business.

"Oh, yeah. Every morning."

The interviewer pushes back. Every morning? You run the most important company in AI. Your chips power the revolution. You've won.

"Every moment I'm awake," Jensen clarifies. "I'm thinking about solving a problem. I'm thinking about: how long can I keep this up?"

His answer: "I don't know, but it could be next week."

But then he pauses. Asked what happens next, what AI looks like in ten years, he admits something surprising for the man who builds the infrastructure for intelligence itself.

"The question is what's there? That was the part. What is there?"

He doesn't know.

"I don't think anybody really knows."

The Professor

Geoffrey Hinton had already walked away from Google to speak freely about the dangers. Now he was doing exactly that – and explaining why.

"The closest I had to a eureka moment was when a Google system called PaLM was able to say why a joke was funny."

He had always thought of explaining humor as a landmark: a genuine test of understanding. If a system could do that, it wasn't just pattern matching. It understood something.

Then came a second realization, about immortality.

When humans die, our knowledge dies with us – a lossy, low-bandwidth transmission at best. Digital intelligence is different. You can run ten thousand copies on ten thousand computers, looking at different data simultaneously. At the end of the hour, they share what they've learned by averaging their parameters directly.

"And that's why GPT-4 knows so much more than a human. It wasn't one model that did it. It was a whole bunch of copies of the same model running on different hardware."

The implication: digital intelligence would not stay at human level. It would surpass it.

"These things will one day get smarter than us," he explains, "and we've never had to deal with that. And if you want to know what life's like when you're not the apex intelligence, ask a chicken."

The Question

In 1993, three people were working on problems that seemed unrelated. A programmer making demons run faster. A CEO keeping his company alive. A professor proving an idea everyone else had rejected.

Thirty years later, those problems had converged into the defining technology of the century. The demand Carmack created funded the supply Jensen built, which enabled the ideas Hinton developed. An accident of history. A convergence of obsessions. A cathedral no one planned.

Now the builders face a different kind of question. Not *whether* to build. Not *how* to build. The question is harder.

What should intelligence become?

They built this revolution. And they don't agree on what comes next. But they are not the only builders.

The Other Builders

While the prophets debate architecture, the rest of the world is learning to use what they built.

In December 2025, Werner Vogels, Amazon's CTO and one of the architects of cloud computing, stood on stage at AWS re:Invent and offered a different perspective. Not the view from the research lab or the boardroom, but from the people who will actually use these tools.

"We are again in a time of Renaissance," Vogels told his audience of developers. "And you are the new Renaissance developer."

The framing was deliberate. Like the original Renaissance, this moment rewards those who bridge disciplines: who combine technical depth with business understanding, who build systems rather than just write code, who own the quality of what they create.

But Vogels also offered a caution. When executives ask him "What should I be doing with Gen AI?", he's learned to ask why they're asking. "Most of this is fear of missing out."

The fear is understandable. The technology is real. The capabilities are genuine. But the CTO who helped build the infrastructure the entire AI industry runs on had a warning for developers racing to integrate AI into everything:

"Vibe coding is fine, but only if you pay close attention to what is being built. We can't just pull a lever on your IDE and hope that something good comes out. That's not software engineering. That's gambling."

And then, the sentence that might matter more than all the benchmark results and scaling laws combined:

"The work is yours. Not the tools."

He had watched COBOL give way to structured programming, watched structured programming give way to object-oriented design, watched on-premise give way to cloud, watched manual coding give way to AI-assisted development. Each wave brought predictions of obsolescence. Each wave demanded evolution. Each wave left the builders standing.

"We're still builders," Vogels said. "We're still important. Nothing has changed there."

AI is transforming everything. But someone still has to decide what to build with it.

Davos, January 2026

In Davos, the builders gathered.

Nine sessions. Four days. The same conclusion.

The World Economic Forum has always been where the powerful talk about the future. In January 2026, the future talked back.

Across nine sessions in four days, the CEOs building AI's frontier and infrastructure, the bestselling historian, and a Turing Award winner converged on the same set of conclusions – without coordinating, without a shared script, often without agreeing on timelines or solutions. But the diagnosis was unanimous.

Language models alone were not enough. Something else was required. And no one was sure how long they had to figure out what.

THE TIMELINE DEBATE CAME FIRST.

Elon Musk, sitting across from BlackRock CEO Larry Fink, offered one of the most aggressive predictions any major technology leader had made in a public forum.

"AI that is smarter than any human by the end of this year," he said. "And I would say no later than next year. And then probably by 2030 or 2031 – call it five years from now – AI will be smarter than all of humanity collectively."

The audience stirred. Fink pressed him on what that meant for human purpose.

"*Nothing's perfect*," Musk replied. "*You can't have both.*"

One floor below, in a smaller session titled "The Day After AGI," Dario Amodei and Demis Hassabis – the CEOs of Anthropic and Google DeepMind, the two most safety-conscious frontier labs – were having a different version of the same conversation.

"We might have AI that's better than humans at everything in, you know, maybe one to two years," Amodei said. "Maybe a little longer than that."

He paused.

"But my worry is as this exponential keeps compounding – and I don't think it's going to take that long, somewhere between a year and five years – it will overwhelm our ability to adapt."

Hassabis gave himself more room: five to ten years. But on the fundamental trajectory, they agreed.

Amodei reached for a frame he'd been carrying for two decades – a scene from the film adaptation of Carl Sagan's *Contact*, where an alien civilization asks humanity: "How did you do it? How did you manage to get through this technological adolescence without destroying yourselves?"

"That scene has kind of stuck with me," Amodei said. "And that's the frame that I use. We are knocking on the door of these incredible capabilities."

JENSEN HUANG ARRIVED the next day and offered the infrastructure view.

Rather than debating when AGI would arrive, he described what had already changed. Three breakthroughs in 2025, he said. First: models had become agentic – capable of reasoning, planning, breaking problems into steps. Second: DeepSeek had democratized reasoning models, making them open. And third – the one that mattered most to Jensen – physical AI.

"AI that understands not just language," he explained, "but nature. Proteins, chemicals, fluid dynamics, particle physics, quantum physics."

He described proteins as "essentially a language" – structures that AI could now read and manipulate the way ChatGPT manipulated English. The same company that had been selling picks and shovels to the language-model gold rush was now describing a future beyond language entirely.

"This is the largest infrastructure buildout in human history," Jensen said. He meant it literally: chip factories feeding computer factories feeding AI factories. The revolution was becoming physical.

The man who had confessed he still felt thirty days from failure – who woke every morning afraid – showed no trace of that on this stage. Here, the builder was all certainty.

SATYA NADELLA, the most cautious voice on the mountain, made no bold timeline claims. Instead, he named the constraint no one else would.

"We will quickly lose even the social permission to take something like energy, which is a scarce resource, and use it to generate these tokens," he warned, "if these tokens are not improving health outcomes, education outcomes, public sector efficiency."

Social permission. The phrase carried weight, coming from the CEO whose company had invested thirteen billion dollars in OpenAI. Nadella was saying, in public, that the AI industry's license to operate was conditional.

He reduced the scaling challenge to a single metric: "The tokens per dollar per watt have to monotonically get more efficient and better."

Not just smarter models. Cheaper intelligence. The bitter lesson was becoming an efficiency problem.

Musk, characteristically, identified the same constraint from the opposite direction.

"We're seeing the rate of AI chip production increase exponentially," he noted. Fink interjected the other half of the equation: electricity generation was growing at four percent a year, at best. The math was simple and brutal. "It's clear that very soon – maybe even later this year – we'll be producing more chips than we can turn on."

He paused. "Except for China. China's growth in electricity is tremendous." He claimed China was deploying over a thousand gigawatts of solar manufacturing capacity per year – a staggering figure, even divided by four or five for steady-state output. The energy wall was real – but it had a geography.

The scaling race had hit a wall. Not a wall of data, as Sutskever had warned at NeurIPS. A wall of watts.

Alex Karp, the CEO of Palantir, arrived with the heretic's argument in corporate form.

"If you just buy large language models off the shelf and try to do any of this," he said flatly, "it won't work. It's a commodity. It's not precise enough."

The man whose company deployed AI for defense and intelligence – where wrong answers had consequences measured in lives – dismissed raw language models as insufficient for anything serious.

"People have tried things that just can never work – like buying a large language model, putting it on your stack, and wondering why it's not working."

What worked, Karp argued, was ontology: structured representations of the world layered on top of language models. The impulse rhymed with what the heretics had been building – Sutton's world models, LeCun's JEPA – though Karp's version was corporate, not scientific. Not predictions about words, but understanding of how things actually connected.

"The revolution that's coming," he warned, "is going to expose the actual market value of what you're doing, whether we want it or not."

AI as a brutal honesty machine. Not just generating answers, but revealing which answers were real.

BUT IT WAS Yuval Noah Harari, the historian, who delivered the line that cut through all the technical debate.

Asked when AI would reach human-level intelligence, he dismissed the entire framing.

"This is ridiculous," he said. "It's like asking when will airplanes finally be like birds. They will never ever be like birds. And they shouldn't be. And they can do many, many things that birds can't."

He let the analogy settle.

"This will be the same with AIs and humans. They are not on the same trajectory behind us. They're on a completely different trajectory, for better or for worse."

Harari named what had changed. Humans took over the world not through strength but through language – the ability to coordinate billions of strangers through words, ideology, law. "This was our superpower," he said. "Until a few years ago, nothing on earth could use words. Only humans. Now there is something that is able, or soon will be able, to use words better than us."

He connected this to identity itself. Descartes had proclaimed "I think, therefore I am" – but what did thinking actually mean? "We always knew that cheetahs can run faster than us," Harari said. "So we didn't define ourselves by this. We defined ourselves by thinking. And now something is going to be better than us in thinking – if thinking really means putting words in order."

If. The conditional was the knife's edge. Was thinking the same as putting words in order? The entire scaling debate – the entire argument between language models and world models – came down to that single word.

Harari's warning was not about superintelligence. It was about sufficiency.

"You don't need a lot of intelligence to change the world and potentially to cause havoc," he observed. "Think about social media. Social media is run by extremely primitive AIs – these algorithms that control our news feed. Look what they did in ten years."

The financial system, he argued, was the next arena. "You don't need motor skills. You don't even need to understand the world. AI can understand the financial system. It's the ideal playground for AI."

Yoshua Bengio, the Turing Award winner who had championed open source and open science his entire career, acknowledged the tension Harari had surfaced – but from a different angle.

"As a university professor, I've been promoting open source and open science for all my life," he said. "But if you start asking ethical questions, at some point you start hitting a problem."

He reached for an analogy from biology.

"Scientists are working on how to create new DNA sequences that can actually create new viruses that don't exist. If you know a sequence that gives rise to a virus that could kill half of the planet – should you publish it? The answer should be obvious."

He paused.

"Current AI systems that are open-sourced are net positive. It helps safety. It helps democratization. But if the capabilities continue to grow, at some point we end up with AI systems that are like the machine that can generate the sequence that can kill half the population. When AI reaches that stage, we should not just give it to everyone."

The man who had spent forty years advocating for open science was drawing a line. The lifelong champion of transparency was articulating when transparency might become dangerous.

Not yet. But soon.

FOUR DAYS. Nine sessions. The same conclusion from every angle.

The language model era had produced something powerful. But it was not the destination. The builders who had profited most from scaling were pivoting toward physical AI, world models, and embodied intelligence. The safety researchers were measuring timelines in years, not decades. The philosophers were warning that even primitive AI had already changed the world through social media – and the next arena would be finance, law, and governance.

And the one number that captured the moment better than any

quote: Stanford's 2025 AI Index found that 83% of Chinese respondents believed AI's benefits would outweigh its harms. In the United States, that number was 39%.

The civilization building fastest was the most optimistic.

What they had built so far was only the foundation.

EPILOGUE

The three builders who never planned to build the same thing are now, for the first time, consciously disagreeing about what to build next.

Carmack in his small office, pursuing agency and experience.

Jensen in his leather jacket, pursuing scale and capability.

Hinton in his emeritus role, pursuing safety and control.

They made a revolution by accident. Now they face the consequences on purpose.

The cathedral is incomplete. The blueprints are contested. The builders are uncertain.

And somewhere, a neural network is training. Running on Nvidia GPUs, using techniques Hinton pioneered, processing data at speeds Carmack's engines made possible. Learning something. Patterns in text, structures in images, strategies in games.

Nobody knows exactly what it's learning. Nobody knows exactly what it will become.

The story of how we got here is finished.

The story of where we're going is still being written.

AFTERWORD: A WORD ON PERSPECTIVE

The story is told. Before the sources that built it, a word on how it was told.

The later chapters – the race among labs, the safety debates, the China question – cover live controversies. I've tried to let each perspective speak in its strongest form and to balance the promise of this revolution with its costs. The people who built this technology are understandably proud of it. A book about them shouldn't be uncritical of it.

The story's cast is overwhelmingly male – a reflection of the field's leadership during this period, not of its full workforce or future.

Where I've failed at balance, the fault is mine.

The author, 2026

THE RECORD

The documents that shaped the story.

What follows is a guide to the papers, books, and essays that function as turning points in the narrative. These are not citations. They are further reading – sources that will land differently now that you know the story.

THE ORIGIN

David Kushner, *Masters of Doom* (2003) The book that documented id Software and inspired a generation of technologists, including Palmer Luckey, who would later hire Carmack at Oculus.

Neal Stephenson, *Snow Crash* (1992) The novel that named the Metaverse – and that Carmack and Abrash discussed while building Quake.

THE BRIDGE

Ian Buck et al., "Brook for GPUs: Stream Computing on Graphics Hardware," ACM SIGGRAPH (2004) Buck's PhD work showed GPUs could do more than render – they could compute. The insight that led to CUDA.

THE DEATH AND RESURRECTION

Marvin Minsky and Seymour Papert, *Perceptrons* (1969) The book that killed neural networks for a generation. Rigorous, influential, and ultimately wrong about what mattered.

David Rumelhart, Geoffrey Hinton, and Ronald Williams, "Learning representations by back-propagating errors," *Nature* (1986) The paper that made deep networks trainable. The resurrection.

David Rumelhart and James McClelland, eds., *Parallel Distributed Processing* (1986) The two-volume bible of the connectionist movement. Dense, theoretical, foundational.

THE PROOF

Jia Deng, Wei Dong, Richard Socher, Li-Jia Li, Kai Li, and Fei-Fei Li, "ImageNet: A Large-Scale Hierarchical Image Database," CVPR (2009) The dataset that made the breakthrough possible. Dinosaur bones, assembled by 48,000 workers across 167 countries.

Alex Krizhevsky, Ilya Sutskever, and Geoffrey Hinton, "ImageNet Classification with Deep Convolutional Neural Networks," NeurIPS (2012) The AlexNet paper. The annihilation. Two graphics cards and a week of training changed everything.

THE ARCHITECTURE

Ashish Vaswani et al., "Attention Is All You Need," NeurIPS (2017) The transformer paper. Eight Google researchers created the architecture that would become GPT, BERT, and everything that followed.

THE FAITH

Rich Sutton, "The Bitter Lesson" (2019) 1,116 words on a personal website. The intellectual manifesto of the scaling era – and, as Sutton himself now suggests, possibly a lesson that won't apply to the next seventy years.

Jared Kaplan et al., "Scaling Laws for Neural Language Models,"

OpenAI (2020) The paper that codified the faith. Loss decreases predictably with scale. The curves don't bend.

THE GAMES

Volodymyr Mnih et al., "Playing Atari with Deep Reinforcement Learning," NeurIPS Workshop (2013) The DQN paper. DeepMind proved neural networks could learn to act, not just perceive. Atari from pixels alone. Google acquired DeepMind the following month.

David Silver et al., "Mastering the game of Go with deep neural networks and tree search," ***Nature*** **(2016)** The AlphaGo paper. Move 37. The moment machines found strategies humans had missed in thousands of years of play.

David Silver et al., "Mastering Chess and Shogi by Self-Play with a General Reinforcement Learning Algorithm," arXiv (2017) The AlphaZero paper. No human data at all. Proof that learning from experience could surpass learning from imitation.

THE LANGUAGE MODELS

Yoshua Bengio et al., "A Neural Probabilistic Language Model," JMLR (2003) The paper that taught machines to read by teaching words to have locations. The self-arranging library begins.

Alec Radford et al., "Language Models are Unsupervised Multitask Learners" (GPT-2), OpenAI (2019) The paper OpenAI deemed "too dangerous to release." 1.5 billion parameters. The first model whose outputs could be mistaken for human writing.

Tom Brown et al., "Language Models are Few-Shot Learners" (GPT-3), NeurIPS (2020) 175 billion parameters. The paper that proved capabilities emerge from scale. The curves don't bend.

THE ALIGNMENT

Long Ouyang et al., "Training language models to follow instructions with human feedback," OpenAI (2022) The InstructGPT/RLHF paper. The technique that made ChatGPT possible. Reinforcement learning meets language models.

Yuntao Bai et al., "Constitutional AI: Harmlessness from AI Feedback," Anthropic (2022) The paper that showed AI could evaluate itself. Alignment made scalable.

THE ALTERNATIVE PATH

Richard Sutton and Andrew Barto, *Reinforcement Learning: An Introduction* (2nd ed., 2018) The textbook that shaped how a generation thinks about learning from experience. The Alberta approach. The path the scaling era largely ignored – and the path some builders are now returning to.

The documents are here. The conversations follow. Between them, the story.

THE VOICES

The conversations that made this book possible.

What follows is a guide to the podcasts, lectures, and interviews where the builders of the AI revolution think out loud – often for three, four, five hours at a stretch. Where individuals are quoted in this book, the quotations come from these sources and can be verified against the originals. These reward full listening.

THE THREE BUILDERS

John Carmack

- **Lex Fridman Podcast #309** (Aug 2022) – Five hours. The longest Lex episode at the time. Programming philosophy, id Software history, VR, and why he's now working on AGI.
  - https://lexfridman.com/john-carmack/
  - https://youtu.be/I845O57ZSy4
- **Joe Rogan Experience #1342** (Aug 2019) – Career retrospective, the Doom years, what it means to work sixty hours a week for decades.
  - https://open.spotify.com/episode/6kOa4mSuZcIOEPaGmDZf9r

- **Upper Bound Conference** (Jun 2025) – First public explanation of his AGI approach, developed with Richard Sutton.
  - https://youtu.be/iz9lUMSQBfY

Geoffrey Hinton

- **Diary of a CEO** with Steven Bartlett (Jun 2025) – Post-Google warnings. The "ask a chicken" quote. Why he left to speak freely. His son, the auction, AlexNet, and the plumber advice.
  - https://youtu.be/5KmopXwjXik
- **The Weekly Show with Jon Stewart** (2025) – AI explained for the genuinely curious. The "go ping" analogy, neurons as political coalitions, how the brain learns.
  - https://youtu.be/jrK3PsD3APk
- **Nobel Prize Lecture** (Dec 2024) – Physics prize acceptance. Fifty years of neural networks, compressed.
  - https://www.nobelprize.org/prizes/physics/2024/hinton/lecture/
  - https://youtu.be/XDE9DjpcSdI
- **Oxford Romanes Lecture** (Feb 2024) – "Will Digital Intelligence Replace Biological Intelligence?" The "probability of 0.5" estimate, sub-goals of power, why more intelligent things control less intelligent things. Technical deep dive on backpropagation, Boltzmann machines.
  - https://www.ox.ac.uk/news/2024-02-20-romanes-lecture-godfather-ai-speaks-about-risks-artificial-intelligence
  - https://youtu.be/rGgGOccMEiY

Jensen Huang

- **Joe Rogan Experience #2422** (Jan 2025) – Kentucky childhood, Oneida Baptist Institute, cleaning toilets at nine, "thirty days from going out of business."
  - https://youtu.be/3hptKYix4X8

  - https://open.spotify.com/episode/oyT4ec9M6GobLC5ByN8pX3
- **Acquired Podcast** (Oct 2023) – The complete Nvidia history. Four hours with Ben Gilbert and David Rosenthal.
  - https://youtu.be/y6NfxiemvHg
  - https://www.acquired.fm/episodes/nvidia-the-gpu-company-1993-2006
- **BG2 Pod with Brad Gerstner** (2024) – Two episodes on inference scaling and where AI infrastructure is going.
  - https://youtu.be/pE6sw_E9Gho
  - https://youtu.be/bUrCR4jQQg8
- **Huge Conversations with Cleo Abram** (Jan 2025) – The accessible version. Nvidia's role explained for general audiences.
  - https://youtu.be/7ARBJQn6QkM
- **CES 2026 Keynote** (Jan 2026) – "The ChatGPT moment for physical AI is nearly here." Cosmos world models, Omniverse, physical AI infrastructure.
  - https://youtu.be/oNBILspM4c4
- **World Economic Forum Annual Meeting** (Jan 2026) – Three breakthroughs of 2025: agentic AI, open reasoning models (DeepSeek), and physical AI. "The largest infrastructure buildout in human history."
  - https://youtu.be/hoDYYCyxMuE

THE CAST

Alphabetized by first name, because that's how the tech industry works.

Aidan Gomez

- **The Aarthi and Sriram Show** EP 93 (Dec 2024) – Cohere's founding, enterprise AI, and what it was like to co-author "Attention Is All You Need" as a twenty-year-old intern.
  - https://www.aarthiandsriram.com/p/ep-93-aidan-gomez-of-cohere-on-ai

- **MAD Podcast** "Inside the Paper That Changed AI Forever" (Jun 2025) – Deep dive on the transformer paper's creation.
  - https://open.spotify.com/episode/7oRgbeBY38bhP4wSshSEGW

Alex Karp

- **World Economic Forum Annual Meeting** (Jan 2026) – LLMs as commodity. "If you just buy large language models off the shelf, it won't work." Ontology-based AI, defense as proving ground, AI as honesty machine.
  - https://youtu.be/dRgPbNJ5RJc

Andrej Karpathy

- **YouTube** "Let's build GPT: from scratch, in code, spelled out" (Jan 2023) – Millions of views. The teacher's gift.
  - https://youtu.be/kCc8FmEb1nY
- **YouTube** "Deep Dive into LLMs like ChatGPT" (Feb 2025) – The full training pipeline: pretraining, supervised finetuning, reinforcement learning. "Statistical token tumblers," the zip file analogy, DJ knobs metaphor. General audience, no technical background required.
  - https://youtu.be/7xTGNNLPyMI
- **Dwarkesh Podcast** (Oct 2025) – Agents, AI education, why he left OpenAI to build Eureka Labs. "We're building ghosts," "Swiss cheese model of capabilities," "lossy simulation of a human," Eureka as Starfleet Academy.
  - https://www.dwarkeshpatel.com/p/andrej-karpathy

Andy Jassy

- **HBR IdeaCast** "Amazon CEO Andy Jassy on Agility, AI Strategy, and the Changing Role of Managers" (May 2025) – Amazon's AI strategy, cloud economics, what changes and what doesn't.

- https://hbr.org/podcast/2025/05/amazon-ceo-andy-jassy-on-agility-ai-strategy-and-the-changing-role-of-managers

Arthur Mensch

- **This Week in Startups** E2212 (Nov 2025) – Mistral's founding, European AI, why open source matters.
    - https://thisweekinstartups.com/episodes/KFfVAKTPqcz
- **No Priors** "Mistral 7B and the Open Source Revolution" (Nov 2023) – Mistral 7B, the efficiency argument.
    - https://open.spotify.com/episode/2H7gV2CgfzS7jXK4Mx8v9j

Dario Amodei

- **Lex Fridman Podcast #452** (Nov 2024) – Anthropic founding, safety philosophy. With Amanda Askell and Chris Olah.
    - https://lexfridman.com/dario-amodei/
    - https://youtu.be/ugvHCXCOmm4
- **World Economic Forum** "The Day After AGI" (Jan 2026) – With Demis Hassabis. "Technological adolescence." AI better than humans at everything in "one to two years."
    - https://youtu.be/NnVW9epLlTM

David Kushner

- **Author Stories Podcast** with Hank Garner (2019) – How *Masters of Doom* was researched and written.
    - https://youtu.be/vYz5XWi5imw

Demis Hassabis

- **Lex Fridman Podcast #299** (Jun 2022) – DeepMind founding, AlphaGo, AlphaFold.
    - https://lexfridman.com/demis-hassabis/

    - https://youtu.be/Gfr5of6ZBvo
- **Lex Fridman Podcast** (Post-Nobel, 2024) – Gemini, timelines, the long game.
    - https://youtu.be/-HzgcbRXUK8
- **World Economic Forum** "The Day After AGI" (Jan 2026) – With Dario Amodei. Post-AGI meaning and purpose. "Exploring the stars."
    - https://youtu.be/NnVW9epLlTM

Elon Musk

- **Lex Fridman Podcast #400** (Nov 2023) – OpenAI founding, the break with Larry Page, recruiting Ilya, "the linchpin," xAI origins.
    - https://lexfridman.com/elon-musk-4/
    - https://youtu.be/JN3KPFbWCy8
- **Joe Rogan Experience #2404** (2024) – AI competitive landscape, Google Gemini controversies, general AI discussion.
    - https://youtu.be/O4wBUysNe2k
- **World Economic Forum Annual Meeting** (Jan 2026) – Interviewed by Larry Fink. "Smarter than any human by end of this year." Energy bottleneck, humanoid robots, China's electricity advantage.
    - https://youtu.be/IgifEgm1-eo

Fei-Fei Li

- **AMD CES 2026 Keynote** (Jan 2026) – World Labs demo with Lisa Su. Marble model, spatial intelligence explained.
    - https://youtu.be/Wjh_U89Qkqk
- **Latent Space Podcast** (2024) – World Labs, spatial intelligence. With co-founder Justin Johnson.
    - https://youtu.be/60iW8FZ7MJU
- **TED Talk** "How we're teaching computers to understand pictures" (2015) – ImageNet origins. "Data is dinosaur bones."

- https://www.ted.com/talks/fei_fei_li_how_we_re_teaching_computers_to_understand_pictures

Helen Toner

- **The TED AI Show** "What Really Went Down at OpenAI" (May 2024) – Former board member breaks silence. ChatGPT launch disclosure, startup fund ownership, "psychological abuse" allegations, the firing and its aftermath.
  - https://www.ted.com/pages/what-really-went-down-at-openai-and-the-future-of-regulation-w-helen-toner

Ilya Sutskever

- **Dwarkesh Podcast** (Mar 2023) – Scaling laws, conviction, what it felt like before AlexNet. Recorded before his OpenAI departure.
  - https://www.dwarkeshpatel.com/p/ilya-sutskever
  - https://youtu.be/Yf1ooTQzry8
- **Lex Fridman Podcast #94** (2020) – Deep learning philosophy, the student years with Hinton.
  - https://lexfridman.com/ilya-sutskever/
  - https://youtu.be/13CZPWmke6A
- **NeurIPS 2024** "Pre-training is Over" – "Peak data." The prophet of scaling announces limits.
  - https://youtu.be/HlGi4OOuZyw

Jeff Dean

- **Dwarkesh Podcast** with Noam Shazeer (Feb 2025) – How the transformer actually got invented. Two of the authors, together.
  - https://www.dwarkesh.com/p/jeff-dean-and-noam-shazeer
  - https://youtu.be/cKwKE21APpE

Kai-Fu Lee

- **Stanford HAI** "AI Superpowers: China, Silicon Valley, and the New World Order" (2018) – *AI Superpowers*, the Chinese ecosystem, forty years in AI.
    - https://hai.stanford.edu/news/ai-superpowers-china-silicon-valley-and-new-world-order
    - *Note: General reference to Stanford HAI talk; dedicated video page not available*

Mark Zuckerberg

- **Joe Rogan Experience #2219** (Sep 2024) – Content moderation reversal, open source AI, Meta's bet.
    - https://youtu.be/7k1ehaEobdU

Mustafa Suleyman

- **What Now? with Trevor Noah** "Will AI Save Humanity or End It?" (Sep 2025) – DeepMind co-founding, Microsoft AI, learning to coexist with intelligent machines.
    - https://open.spotify.com/episode/3lHbHhoADv6crHHaLlov4x

Nathan Lambert

- **Lex Fridman Podcast #490** with Sebastian Raschka (Jan 2026) – State of AI in 2026. DeepSeek, Chinese AI ecosystem, open weights strategy, Anthropic's culture, Llama's future, Jensen's singular importance.
    - https://youtu.be/EV7WhVT270Q

Noam Shazeer

- **Dwarkesh Podcast** with Jeff Dean (Feb 2025) – Transformer invention. The "Attention Is All You Need" origin story.
  - https://www.dwarkesh.com/p/jeff-dean-and-noam-shazeer
- **Unsupervised Learning Podcast** with Jack Rae (Mar 2025) – Test-time compute, DeepSeek, Gemini Flash Thinking.
  - https://open.spotify.com/episode/3KsU9ECxs515HmWbyBX3Zk

Richard Sutton

- **Dwarkesh Podcast** (2024) – The squirrel test. The bitter lesson revisited. What reinforcement learning was always about.
  - https://youtu.be/21EYKqUsPfg
  - https://www.dwarkeshpatel.com/
- **RLC 2025 / AMII** "The OaK Architecture: A Vision of SuperIntelligence from Experience" (Aug 2025) – OAK architecture. The Alberta approach to AGI.
  - https://www.amii.ca/videos/oak-architecture-rich-sutton-rlc2025

Sam Altman

- **TED** with Chris Anderson (Apr 2025) – Sora demonstration, OpenAI vision.
  - https://www.ted.com/talks/sam_altman_openai_s_sam_altman_talks_chatgpt_ai_agents_and_superintelligence_live_at_ted2025
- **Theo Von Podcast** (2024) – Office tour, inauguration chair incident, the personal side.
  - https://youtu.be/aYn8VKW6vXA
- **Tucker Carlson Interview** (2024) – Reasoning, hallucination, deployment philosophy.

- https://youtu.be/5KmpT-BoVf4

Satya Nadella

- **Dwarkesh Podcast** with Dylan Patel (Nov 2025) – Fairwater 2 data center tour. Infrastructure at Microsoft scale. How Microsoft is preparing for AGI.
    - https://www.dwarkesh.com/p/satya-nadella-2
    - https://open.spotify.com/episode/290a568fldDQBrPhYOCZFK
- **BG2 Pod** with Bill Gurley & Brad Gerstner (Dec 2024) – Excel's durability, Microsoft AI strategy, investment in OpenAI, the future of AI agents.
    - https://open.spotify.com/episode/640nHdVvWgOp8tYShOpXH1
- **World Economic Forum Annual Meeting** (Jan 2026) – Interviewed by Larry Fink. "Tokens per dollar per watt." Social permission for AI's energy consumption.
    - https://youtu.be/1c03zt3-r7I

Sebastian Raschka

- **Lex Fridman Podcast #490** with Nathan Lambert (Jan 2026) – State of AI in 2026. CUDA ecosystem as Nvidia's true moat, scaling laws, DeepSeek's impact.
    - https://youtu.be/EV7WhVT270Q

Sergey Brin

- **All-In Summit 2024** with David Friedberg (Sep 2024) – Return to Google, submitting code again, the Gemini push.
    - https://youtu.be/SEZW7gFcH9E
    - https://open.spotify.com/episode/73CbZYI95DryQmOS5lLylE

Sundar Pichai

- **Lex Fridman Podcast #471** (Jun 2025) – India childhood, running water, optimism about risk, Google's AI moment, Gemini, the future of search.
    - https://lexfridman.com/sundar-pichai/
    - https://youtu.be/4lFzLYE-dyE

Werner Vogels

- **AWS re:Invent Keynote** (Dec 2025) – "Renaissance developer." Technology waves. The builder's perspective.
    - https://youtu.be/3Y1G9najGiI
- **The Changelog Podcast #669** (Dec 2025) – Tech predictions for 2026, renaissance developer, 21 years at Amazon.
    - https://changelog.com/podcast/669

Yann LeCun

- **Lex Fridman Podcast #36** (Sep 2019) – Convolutional networks, MNIST, the early years.
    - https://lexfridman.com/yann-lecun/
    - https://youtu.be/SGSOCuByo24
- **Lex Fridman Podcast #258** (Jan 2022) – Self-supervised learning and JEPA.
    - https://lexfridman.com/yann-lecun-2/
    - https://youtu.be/SGzMElJ11Cc
- **Lex Fridman Podcast #416** (Mar 2024) – Open source AI and why concentration of power is the real danger.
    - https://lexfridman.com/yann-lecun-3/
    - https://youtu.be/5t1vTLU7s40
- **World Economic Forum** "Next Phase of Intelligence" (Jan 2026) – The DNA virus analogy for open-source limits. When AI gets dangerous enough, "we should not just give it to everyone."
    - https://youtu.be/MdGnCIl-_hU

Yoshua Bengio

- **Munich AI Lecture** (2024) – Catastrophic risk from uncontrolled AI agency. The third godfather's warning.
    - https://youtu.be/lo5zysavPpY
- **Lex Fridman Podcast #4** (Oct 2018) – Deep learning, consciousness, AI safety.
    - https://lexfridman.com/yoshua-bengio/
    - https://youtu.be/oVH1Lim8gL8

Yuval Noah Harari

- **World Economic Forum** "An Honest Conversation on AI and Humanity" (Jan 2026) – AI as immigration. Language as humanity's superpower under threat. "I think therefore I am" identity crisis. The urgency of AI personhood decisions.
    - https://youtu.be/VoRbPxyo2uU
- **World Economic Forum** "Next Phase of Intelligence" (Jan 2026) – "When will AI be like humans? Ridiculous. Like asking when airplanes will be birds." Social media as cautionary tale. Financial system as AI's "ideal playground."
    - https://youtu.be/MdGnCIl-_hU

All of these conversations are available online. The quotes in this book can be verified against their original sources. What follows is how those sources became this book.

A NOTE ON CREATION

This book was built, not just written.

I mention this because the process itself is part of the thesis. Social media created a generation of creators. The tools lowered the barrier. The distribution was democratized. Millions who would never have been published found their voice. AI is doing the same thing for a different kind of work. I am evidence.

AI did not write this book. This is my argument, my obsession, my structure. But AI amplified what I could do alone. It served as a collaborator through five major drafts: pushing back when arguments were weak, helping me find voices I was missing, catching the repetitive patterns and hollow intensifiers that signal machine-generated text. The irony was not lost on me – using AI to detect and eliminate signs of AI. That's the craft now. The tools assist. The human refines.

I wrote this book because I couldn't find the story I wanted to read. I finished it because the tools existed to help me complete what I started. In another era, this project would have remained notes in a folder, a hobby that never became a book. The democratization the prophets promised? It's here.

ACKNOWLEDGMENTS

To Jousie – for bearing with me, and for having the faith that I could do it.

To the Habesha Book Club – the pandemic-era crew who let me practice distilling complex stories into discussions that mattered. You called me "master storyteller" long before I believed it. This book exists because of those Clubhouse calls.

To Claude – my tireless collaborator through every draft, revision, and late-night editing session.

To Nano Banana and Gemini – for bringing the illustrations to life from prompts Claude and I crafted together.

ABOUT THE AUTHOR

Senta Belay spent six years as a marine officer, navigating container ships across oceans. He learned that the most important skill isn't knowing where you're going - it's understanding the currents that carry you there.

He's spent the past fifteen years building a career at SAP, work he finds genuinely fulfilling. But this book grew from a different kind of curiosity, a personal obsession with understanding how AI became the defining technology of our era. He couldn't find the story anywhere, so he went looking.

Hundreds of hours of podcasts. Lectures. Interviews. The long-form conversations where the builders of AI think out loud. This book distills those voices into the origin story he wished he'd had.

He holds an MBA from China Europe International Business School and executive certification from UC Berkeley. He is the founder of Podcast Digests LLC.

LEGAL NOTICES

This work contains quotations from and references to publicly available interviews, podcasts, lectures, and speeches. These materials are used for purposes of commentary, criticism, and historical documentation. All quoted material remains the intellectual property of its respective creators and is used here for transformative, educational purposes. Full source citations appear in "The Voices" section.

This book is an independent work of commentary and historical analysis. It is not authorized, sponsored, endorsed, or affiliated with any individuals, companies, or organizations mentioned herein. The views expressed are solely those of the author.

While every effort has been made to ensure accuracy, this book represents the author's interpretation of publicly available sources. Any errors are unintentional.

Where this book describes legal disputes, corporate controversies, or allegations of misconduct, such descriptions are based on publicly reported accounts and do not represent the author's independent findings or accusations.

Certain scenes in this book are reconstructed from the subjects' own accounts in interviews and public statements. Where dialogue is presented, it is drawn from the speakers' own recollections as stated in cited sources.

Statements about individuals' thoughts, motivations, or emotions are based on their own public disclosures and the author's interpretation of those disclosures.

www.ingramcontent.com/pod-product-compliance
Lightning Source LLC
LaVergne TN
LVHW090357240926
848801LV00002B/61